Mathematics Galore!

The First Five Years of the St. Mark's Institute of Mathematics

© 2012 by

The Mathematical Association of America (Incorporated)

Library of Congress Catalog Card Number 2012937494

Print edition ISBN 978-0-88385-776-2

Electronic edition ISBN 978-1-61444-103-8

Printed in the United States of America

Current Printing (last digit):
10 9 8 7 6 5 4 3 2 1

Mathematics Galore!

The First Five Years of the St. Mark's Institute of Mathematics

James Tanton

 MAA

Published and Distributed by
The Mathematical Association of America

CLASSROOM RESOURCE MATERIALS

Classroom Resource Materials is intended to provide supplementary classroom material for students—laboratory exercises, projects, historical information, textbooks with unusual approaches for presenting mathematical ideas, career information, etc.

101 Careers in Mathematics, 2nd edition edited by Andrew Sterrett

Archimedes: What Did He Do Besides Cry Eureka?, Sherman Stein

Calculus: An Active Approach with Projects, Stephen Hilbert, Diane Driscoll Schwartz, Stan Seltzer, John Maceli, and Eric Robinson

The Calculus Collection: A Resource for AP and Beyond, edited by Caren L. Diefenderfer and Roger B. Nelsen

Calculus Mysteries and Thrillers, R. Grant Woods

Conjecture and Proof, Miklòs Laczkovich

Counterexamples in Calculus, Sergiy Klymchuk

Creative Mathematics, H. S. Wall

Environmental Mathematics in the Classroom, edited by B. A. Fusaro and P. C. Kenschaft

Excursions in Classical Analysis: Pathways to Advanced Problem Solving and Undergraduate Research, by Hongwei Chen

Exploratory Examples for Real Analysis, Joanne E. Snow and Kirk E. Weller

Geometry From Africa: Mathematical and Educational Explorations, Paulus Gerdes

Historical Modules for the Teaching and Learning of Mathematics (CD), edited by Victor Katz and Karen Dee Michalowicz

Identification Numbers and Check Digit Schemes, Joseph Kirtland

Interdisciplinary Lively Application Projects, edited by Chris Arney

Inverse Problems: Activities for Undergraduates, Charles W. Groetsch

Keeping it R.E.A.L.: Research Experiences for all Learners, Carla Martin and Anthony Tongen

Laboratory Experiences in Group Theory, Ellen Maycock Parker

Learn from the Masters, Frank Swetz, John Fauvel, Otto Bekken, Bengt Johansson, and Victor Katz

Math Made Visual: Creating Images for Understanding Mathematics, Claudi Alsina and Roger B. Nelsen

Mathematics Galore!: The First Five Years of the St. Marks Institute of Mathematics, James Tanton

Ordinary Differential Equations: A Brief Eclectic Tour, David A. Sànchez

MAA Service Center
P.O. Box 91112
Washington, DC 20090-1112
1-800-331-1MAA FAX: 1-301-206-9789

The true joy in mathematics, the true hook that compels mathematicians to devote their careers to the subject, comes from a sense of boundless wonder induced by the subject. There is transcendental beauty, there are deep and intriguing connections, there are surprises and rewards, and there is play and creativity. Mathematics has very little to do with crunching numbers. Mathematics is a landscape of ideas and wonders.

This book is for my students, who have come to understand what I mean by this.

Contents

Appendices

Indexes

Introduction

Eight years ago I took the plunge: I left the college and university world and became a high school mathematics teacher. It was time for me to do what I preached.

For many years I conducted workshops for educators that I thought were innovative, exciting and spoke directly to the mathematical experience. My goal was to enhance joyful learning for students through joyful work with their teachers and I designed activities that illustrated, I hoped, the value of intellectual play, of discovering and owning ideas, of questioning assumptions, of flailing, and of experiencing success. Although deemed interesting and fun my workshops, however, were rarely seen by my colleagues as relevant to the actual classroom experience. They weren't directly "curriculum focused."

When I consulted with teachers about this they spoke simply of their lack of time to "get through" the material assigned, it being a practical reality to push content over process. It is easier, even necessary many said, to present jargon first and establish context later, to practice mechanics over understanding, and to rely on rote thought rather than adaptability of thought. The mindset of standardized testing and college-board curricular, they felt, encourages this and I heard that there is little or no room for creative exploration in the high school classroom. And a number of teachers confessed to not knowing what innovative thinking in mathematics could mean. My workshops did not speak to their experience and their perceived needs.

I believe that the middle and high school years are formative years. Students at this time are engaging with the intellectual world, thinking big questions and playing with ideas. They are naturally flexible thinkers. The idea of promoting rote thinking over playful thought deeply perturbs me.

As a mathematician I want students to see the subject I love and experience its creative art. But I also believe that teaching mathematics is essential for a broad and important reason: true innovation and progress in business and in research comes from adaptable thinking, the type that true mathematics encourages and refines. I believe that mindful, flexible thinking is a vital 21st-century skill we need to teach. The push for content in mathematics shouldn't squelch creativity.

Seven years ago I began to wonder if it is possible to teach flexibility of thought and practice joyful learning in the secondary world. Can one teach the quadratic formula through a process of artful discovery, even if the exam is next Thursday? Could I do what I was asking teachers to do? I decided I had to try. In fact, I felt morally obligated to try.

In 2004 I joined the faculty of St. Mark's School in Southborough, Massachusetts, and became a high school mathematics teacher.

The St. Mark's Institute of Mathematics

I am grateful to Tony and Elsa Hill, former Heads of School, who saw in my work and in my goals an opportunity for St. Mark's School. They offered me not only the chance to think deeply about the secondary mathematics curriculum and to teach it, but also to found a new Institute of Mathematics, a concrete point of intersection between the secondary and college worlds. I couldn't refuse!

The St. Mark's Institute of Mathematics, now in its seventh full year, has a simple mission:

- To provide community outreach of mathematical excellence; and
- To enhance and promote creative mathematical thinking, awareness, and enjoyment of the subject.

I meet these goals by offering workshops, courses and activities for students (both from St. Mark's School and from the wider Boston community), professional development and graduate courses for mathematics teachers (with accreditation from Northeastern University), and public lectures, activities and written materials for the general public.

Three Institute products are particularly popular: the mathematics research classes for students, the weekly e-mail puzzles, and the monthly mathematical newsletters. This book showcases mathematical output from these three activities.

The student research classes meet for one hour a week for 10 sessions throughout a semester. Their goal is to engage in sustained thinking on a single research problem, and fully experience the frustrations and joys of the research process.

Students attending the research classes have engaged in original mathematics and have co-authored expository and research articles. They have discovered, for example, a novel approach to the Borsuk-Ulam theorem, they have found a visual proof of the Galilean ratios that extends the result to ratios of even more fractions, and they have proved Pick's theorem in a new way. (Details appear in Appendices II, IV, and V.) As you read the material of this book you will learn of the wonderful creative thinking young folk can offer. I cannot overstate how proud I am of my students.

Although I avoid any sense of competition in my work with students, I do send out a weekly e-mail puzzle and offer a prize (usually one of "Dr. T.'s should-be-world-famous double-chocolate coconut fudge brownies"). I make the point that I seek succinct and beautiful solutions, hoping to encourage refinement of thought and an awareness of elegance. Close to 700 people are currently on the Institute e-mail puzzle list. Some past puzzles appear in this book along with student thoughts about them.

The mathematical newsletters have proved to be tremendously popular. A sample of 26 newsletters is the base of this book.

The Newsletters

Each month I produce a hard-copy newsletter to send out across the globe. There are about 650 recipients to date. The popularity of the Institute is solely through word of mouth and I have found that producing tangible letters has helped considerably in this regard: copies left behind in departmental common rooms, in teacher lounges, and in the backs of airplane seats(!) have brought on board new readers.

As you read the pieces in this book you may think that the title "newsletter" is a little odd. The letters here originally contained announcements and advertisements for upcoming Institute workshops and activities and so in that sense offered news. I have removed those announcements and only the mathematical content remains. I have also edited the letters. (The life of a high-school teacher is one of extreme busyness and not every piece was written with absolute care before being sent!)

The richness of this book comes from the essays that follow each newsletter. In them I delve more deeply into the mathematics mentioned in the letter, I describe student and teacher work that I received in response to the newsletter, and I offer thoughts on how to push ideas in new directions, along with questions and mysteries for further exploration.

Who this Book is For

This book is for all who wish to explore mathematics!

Teachers

I am fully aware that the content of this book does not directly address the issues I raise at the beginning of this introduction. For high-school teachers this book, at face value, is for extra-curricular use (though a number of sections are directly related to the standard school curriculum). However, this book sits in balance with a slew of curricular material I have produced, self-published and posted on my personal website, www.jamestanton.com, all of which is my direct answer to the challenge of bringing joyful thinking into standard mathematics teaching. This book and that content give a complete picture of my thoughts on mathematical thinking and mathematics teaching.

On another level, high-school teachers have already found direct value in my newsletters and will be sure to find further depth and value in the essays that follow them here. I suggest that any teacher perusing sections of this book for personal enjoyment should still hold an active mind as a teacher. This will allow for cross-pollination of ideas. Many of these newsletters are inspired by my work with students in the regular classroom and many teachers have described to me moments of spark in their teaching that have come from thinking about a newsletter.

It is also eye-opening to many teachers to see in this book the mathematical heights to which young minds can soar. This is an important reason for teachers to read this book.

Student Teachers

This book provokes reflection on the high school mathematical experience from a college-mathematics perspective and so is in direct alignment with the mission of students in teacher preparation courses. A goal of this book is to sustain enjoyment of mathematics and to offer insights on how to share that enjoyment with younger students. (It is also just a good read for anyone who likes mathematics!)

Math Club Organizers

Each newsletter offers rich, but accessible, mathematics with puzzles and curiosities to provide lively and interesting math club topics. The essays that follow show how to delve deeply into the mathematics.

Faculty

The material of this book could be used as the basis of a general content-oriented prospective teachers' course or a mathematics appreciation course. Number theory, geometry, combinatorics, probability, and calculus all make their appearances in ways that students find surprising and exciting. Faculty looking for fresh ideas to insert into an existing course will find this book useful too. The indexes at the end of the book make finding relevant tidbits easy.

Mathematicians

I am a mathematician and I am intrigued by mathematics. The list of "Classic theorems proved" at the end of the book shows that significant mathematics is discussed. There are plenty of intriguing problems sprinkled throughout the text along with many suggestions for continued research and study.

Math Majors and Math Enthusiasts

The St. Mark's Institute of Mathematics has followers from many different backgrounds who enjoy mathematics. This book has value for all who like math.

Math Circles

Extracurricular programs called Math Circles are growing in popularity in the U.S., and my work and this book are directly relevant to math circle activities. See appendices II, III, IV and V in particular for a sense of my approach to circle work. Each newsletter stands as a good topic for a math circle session.

To learn about Math Circles see:

SIGMAA MCST: http://sigmaa.maa.org

This is a Special Interest Group supported by the Mathematical Association of America on Math Circles for Students and Teachers.

AIM Math Teachers' Circle: www.mathteacherscircle.org
This is a program supported by the American Institute of Mathematics.

MSRI National Association of Math Circles: www.mathcircles.org
This is a program supported by the Mathematical Science Research Institute.

How to Use this Book

Organizing this book provided a conundrum. Many newsletters touch on multiple topics, essays vary in their level of mathematical sophistication, and I have included additional essays and articles produced by the Institute. As there is no clear system for categorizing the pieces I decided to simply list the newsletters in alphabetical order by their titles and set the additional materials as appendices.

The best way to read this book for personal pleasure is to pick essays and articles in no particular order and enjoy the mathematical stories that unfold. Each newsletter and essay that follows it is a self-contained pair.

For those seeking specific materials to include in a course or a project, the indexes at the back of the book will help.

- Index I on p. 257 lists broad mathematical topics and the articles that touch on them.
- Index II on p. 259 lists classic theorems that are proved in this book and the location of those proofs.

About St. Mark's School

Located in Southborough, Massachusetts, St. Mark's School is a co-educational, college preparatory boarding school affiliated with the Episcopal Church. The School was founded in 1865 and currently enrolls 335 boarding and day students from 21 states and 14 countries.

St. Mark's School supports and values the work of the St. Mark's Institute of Mathematics.

For more information about the School and the Institute, see www.stmarksschool.org.

Acknowledgements

My deep gratitude to Don Albers and David Auckly for encouraging me to do this project, and to Underwood Dudley for deleting all but one appearance of "notice that" from the original manuscript (and offering many other invaluable editorial comments).

Arctangents

PUZZLER: Stacking Right Triangles

The following diagram starts with a right isosceles triangle with legs 1 and stacks an additional right triangle with a leg of 1 onto the hypotenuse of a previously constructed right triangle.

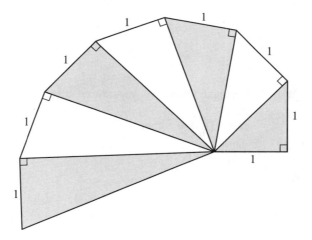

a) If n right triangles are stacked in this way, what is the length of the longest line segment in the diagram?

b) If we keep stacking right triangles, will the diagram ever make a full turn of rotation? Two full turns of rotation?

TIDBIT: Stacking Slopes

The angle a line of slope m makes with the horizontal is called the *arctangent* of the slope, denoted $\arctan(m)$. For example, a line of slope 1 makes a 45° angle with the horizontal so

$$\arctan(1) = 45°$$

The following line of slope $\frac{1}{2}$ makes an angle x with the horizontal:

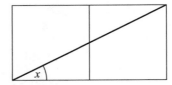

It turns out x is approximately $26.6°$, so

$$\arctan\left(\frac{1}{2}\right) \approx 26.6°.$$

Turning the diagram sideways gives a line of slope 2 with angle to the horizontal $y = 90 - 26.6 = 63.4°$. (We will ignore the negative slope.)

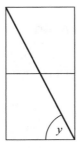

We have $\arctan(2) \approx 63.4°$

In general, looking at the following diagram two ways we see that:

$$\arctan\left(\frac{a}{b}\right) + \arctan\left(\frac{b}{a}\right) = 90°$$

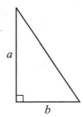

If we look at other diagrams in clever ways we can discover many curious relations between angles of different slopes. For example:

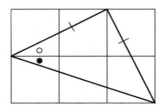

$$\arctan\left(\frac{1}{2}\right) + \arctan\left(\frac{1}{3}\right) = 45°$$

(The angle marked with an open circle sits with a line of slope $1/2$ and the angle with a closed circle a line of slope $1/3$. The two angles add to one of the base angles of an isosceles triangle, which contains a right angle. Thus the base angle is actually $45°$.)

Consider this picture:

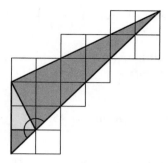

Do you see that:

$$\arctan(1) + \arctan(2) + \arctan(3) = 180°?$$

[The previous two diagrams are due to Ed Harris.]

This picture

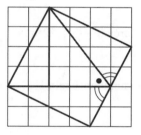

shows that

$$2\arctan(2) + \arctan\left(\frac{4}{3}\right) = 180°.$$

Here is a more complicated picture:

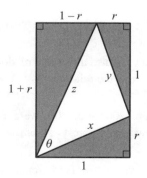

Pythagoras's Theorem and some algebra show that $x = \sqrt{1+r^2} = y$ and $z = \sqrt{(1-r)^2 + (1+r)^2} = \sqrt{2+2r^2} = \sqrt{2} \cdot x$. This establishes that the central triangle is a 45-90-45 triangle. Thus $\theta = 45°$ and

$$\arctan(r) + \arctan\left(\frac{1-r}{1+r}\right) = 45°.$$

(Put in $r = 1/2$ to see a familiar result.)

Here are more intriguing angle tidbits:

$$2\arctan\frac{1}{2} - \arctan\frac{1}{7} = 45° \qquad\qquad \text{(Herman, ca. 1706)}$$

$$4\arctan\frac{1}{5} - \arctan\frac{1}{70} + \arctan\frac{1}{99} = 45° \qquad\qquad \text{(Euler, 1738)}$$

$$2\arctan\frac{1}{3} + \arctan\frac{1}{7} = 45° \qquad\qquad \text{(Hutton, ca. 1778)}$$

$$12\arctan\frac{1}{18} + 8\arctan\frac{1}{57} - 5\arctan\frac{1}{239} = 45° \qquad\qquad \text{(Gauss, Date unclear)}$$

$$12\arctan\frac{1}{49} + 32\arctan\frac{1}{57} - 5\arctan\frac{1}{239} + 12\arctan\frac{1}{110443} = 45° \qquad \text{(Takano, 1982)}$$

Fibonacci Surprise. The Fibonacci numbers begin $1, 1, 2, 3, 5, 8, 13, 21, 34, \ldots.$

We have

$$\arctan 1 = \arctan\frac{1}{2} + \arctan\frac{1}{3}$$

$$\arctan\frac{1}{3} = \arctan\frac{1}{5} + \arctan\frac{1}{8}$$

$$\arctan\frac{1}{8} = \arctan\frac{1}{13} + \arctan\frac{1}{21}$$

$$\arctan\frac{1}{21} = \arctan\frac{1}{34} + \arctan\frac{1}{55}$$

and so on! The first relation, for instance, states that a line of slope $1/2$ stacked with a line of slope $1/3$ gives a line of slope 1. (Can you prove the relations?)

Internet Research. Scholars of the 1700s and 1800s were interested in computing the digits of π. Research why formulas like these were of interest to them.

Research Corner. Create your own visual proofs of arctangent discoveries.

COMMENTARY, SOLUTIONS and THOUGHTS

As mentioned at the end of the newsletter the motivation for developing arctangent formulas was computing digits of π. (See [BLATNER], [KNOTT], for instance.) Allow me to explain.

Warning. This portion of the essay uses calculus.

Begin with the geometric series

$$\frac{1}{1-x} = 1 + x + x^2 + x^3 + \cdots$$

and substitute $-x^2$ for x. This gives

$$\frac{1}{1+x^2} = 1 - x^2 + x^4 - x^6 + \cdots$$

Integrating term by term (assuming integration commutes with summation) we have

$$\arctan x = x - \frac{1}{3}x^3 + \frac{1}{5}x^5 - \frac{1}{7}x^7 + \cdots$$

known as Gregory's Series (in honor of astronomer and algebraist James Gregory, 1638–75). (See [ROY].) It is valid for $-1 \le x \le 1$ provided angles are measured in radians rather than degrees.

Putting in $x = 1$ gives

$$\frac{\pi}{4} = 1 - \frac{1}{3} + \frac{1}{5} - \frac{1}{7} + \frac{1}{9} - \cdots$$

also sometimes referred to as Gregory's Series but more commonly known as Leibniz's Series.

The series gives us a way of computing $\frac{\pi}{4}$, but it converges very slowly.

Student Project. Have students use the series to compute the value of π to one decimal place. Next have them attempt to compute π accurately to two decimal places.

The sixth term of the series, $1/11$, is the first term that has a zero after the decimal point when expressed as a decimal, the 51st term $1/101$ is the first term with two zeros after the decimal point, and so on. This shows that at the least it will take more than half a million terms to determine the value of $\frac{\pi}{4}$ to six decimal places.

To get a faster rate of convergence for $\arctan x = x - \frac{1}{3}x^3 + \frac{1}{5}x^5 - \frac{1}{7}x^7 + \cdots$ we could use a value of x smaller than 1. For example, choosing $x = \frac{1}{\sqrt{3}}$ gives

$$\frac{\pi}{6} = \frac{1}{\sqrt{3}} - \frac{1}{3} \cdot \frac{1}{\sqrt{3}^3} + \frac{1}{5} \cdot \frac{1}{\sqrt{3}^5} - \cdots$$

or

$$\pi = 2\sqrt{3}\left(1 - \frac{1}{3 \cdot 3^1} + \frac{1}{5 \cdot 3^2} - \frac{1}{7 \cdot 3^3} + \frac{1}{9 \cdot 3^4} - \cdots\right)$$

The 13th term of the series, when expressed as a decimal, has six zeros after the decimal point, showing that this series converges significantly faster.

Student Project. Using only the tools available in the 1700s (namely, pencil and paper!) use the series to compute the value of π to some degree of accuracy. Is it easy to work with the series?

In using $\arctan x = x - \frac{1}{3}x^3 + \frac{1}{5}x^5 - \frac{1}{7}x^7 + \cdots$ it would be easier to work with a value x that is compatible with our base-10 system of arithmetic, for example, with $x = \frac{1}{10}$, $x = \frac{1}{5}$, or $x = \frac{1}{20}$. The trouble with such values, however, is that their arctangents are not known. (What is $\arctan\left(\frac{1}{10}\right)$?)

In 1706, John Machin ([MACTUTOR]) discovered a remarkable arctangent formula that involves $\frac{1}{5}$. It allowed him to compute by hand the first 100 digits of π, an astonishing accomplishment! He used

$$\frac{\pi}{4} = 4\arctan\frac{1}{5} - \arctan\frac{1}{239}.$$

Although the number 239 is awkward, the powers of $\frac{1}{239}$ become small quickly and only the first few terms of the series for $\arctan\left(\frac{1}{239}\right)$ need be computed to obtain a high degree of accuracy.

Challenge. Verify Machin's formula by computing $(5+i)^4(-239+i)$. Compare the arguments of $5+i$, $-239+i$ and $(5+i)^4(-239+i)$. (Or create a visual proof of Machin's result akin to the methods of the newsletter?)

Machin's success inspired other mathematicians to seek useful arctangent formulas. We presented some in the newsletter.

The Fibonacci Numbers

Let F_n denote the n th Fibonacci number with $F_0 = 1$, $F_1 = 1$, and $F_{n+2} = F_{n+1} + F_n$ for $n \geq 1$. In the newsletter we claim

$$\arctan\left(\frac{1}{F_{2n-1}}\right) = \arctan\left(\frac{1}{F_{2n}}\right) + \arctan\left(\frac{1}{F_{2n+1}}\right)$$

for $n \geq 1$. Using $\tan(\alpha + \beta) = \frac{\tan\alpha + \tan\beta}{1 - \tan\alpha \tan\beta}$ we see

$$\tan\left(\arctan\left(\frac{1}{F_{2n}}\right) + \arctan\left(\frac{1}{F_{2n+1}}\right)\right) = \frac{\frac{1}{F_{2n}} + \frac{1}{F_{2n+1}}}{1 - \frac{1}{F_{2n}F_{2n+1}}} = \frac{F_{2n} + F_{2n+1}}{F_{2n}F_{2n+1} - 1}.$$

We will show that the right-hand side of this equation equals $\frac{1}{F_{2n-1}}$.

Cassini's identity $F_n^2 - F_{n-1}F_{n+1} = (-1)^n$ (easily established by induction) gives $-1 = F_{2n-1}F_{2n+1} - F_{2n}^2$. Substituting this value into the denominator of the expression on the right of the above equation we obtain

$$\frac{F_{2n} + F_{2n+1}}{F_{2n}F_{2n+1} + \left(F_{2n-1}F_{2n+1} - F_{2n}^2\right)} = \frac{F_{2n} + F_{2n+1}}{(F_{2n} + F_{2n-1})F_{2n+1} - F_{2n}^2}$$

$$= \frac{F_{2n} + F_{2n+1}}{F_{2n+1}^2 - F_{2n}^2}$$

$$= \frac{1}{F_{2n+1} - F_{2n}}$$

$$= \frac{1}{F_{2n-1}},$$

as hoped. (Also see [KALMAN and MENA] for more on this.)

If we string together the relations

$$\arctan 1 = \arctan \frac{1}{2} + \arctan \frac{1}{3},$$

$$\arctan \frac{1}{3} = \arctan \frac{1}{5} + \arctan \frac{1}{8},$$

$$\arctan \frac{1}{8} = \arctan \frac{1}{13} + \arctan \frac{1}{21},$$

$$\arctan \frac{1}{21} = \arctan \frac{1}{34} + \arctan \frac{1}{55},$$

we obtain the following curious infinite sum:

$$\arctan (1) = \arctan \left(\frac{1}{2}\right) + \arctan \left(\frac{1}{5}\right) + \arctan \left(\frac{1}{13}\right) + \arctan \left(\frac{1}{34}\right) + \cdots$$

which can be written as:

$$\arctan \left(\frac{1}{F_0}\right) = \arctan \left(\frac{1}{F_2}\right) + \arctan \left(\frac{1}{F_4}\right) + \arctan \left(\frac{1}{F_6}\right) + \cdots$$

Question. In 2003, Ko Hayashi [HAYASHI] presented visual proofs of the relations $\arctan \left(\frac{1}{F_{2n-1}}\right) = \arctan \left(\frac{1}{F_{2n}}\right) + \arctan \left(\frac{1}{F_{2n+1}}\right)$. Is there an analogous visual demonstration of this infinite sum?

A Note on the Decomposition of $\frac{\pi}{4}$

In the newsletter we established

$$\frac{\pi}{4} = \arctan (r) + \arctan \left(\frac{1-r}{1+r}\right)$$

for $0 < r < 1$. It is interesting to note that *any* decomposition of $\frac{\pi}{4}$ as a sum of two arctangents must be of this form! To see this, suppose that $\frac{\pi}{4} = \arctan r + \arctan s$ with $0 < r, s < 1$. Then

$$1 = \tan \left(\frac{\pi}{4}\right) = \frac{\tan (\arctan r) + \tan (\arctan s)}{1 - \tan (\arctan r) \tan (\arctan s)} = \frac{r + s}{1 - rs},$$

yielding $s = \frac{1-r}{1+r}$.

Comment. For more on the decomposition of angles into sums of arctangents see [KOWALSKI] and [TODD].

Other Identities

The results in the newsletter are based on slopes of lines drawn in square grids. We need not be restricted in this way! In April of 2009, Bob Cornell created a visual result based on a regular hexagon.

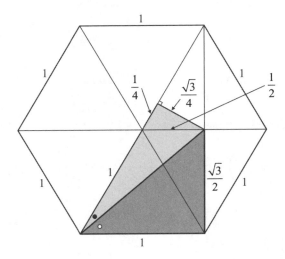

$$\frac{\pi}{3} = \arctan\left(\frac{\sqrt{3}}{2}\right) + \arctan\left(\frac{\sqrt{3}}{5}\right)$$

The Spiral of Theodorus

The opening puzzler describes the Spiral of Theodorus named after Theodorus of Cyrene, circa 450 B.C.E. (See [DAVIS], [GRONAU] and [HAHN and SCHOENBERGER], for example.) Repeated application of Pythagoras's theorem shows that the length of the hypotenuse of the n th triangle in the spiral is $\sqrt{n+1}$.

A stack of n triangles turns through an angle

$$\varphi_n = \arctan(1) + \arctan\left(\frac{1}{\sqrt{2}}\right) + \cdots + \arctan\left(\frac{1}{\sqrt{n}}\right)$$

and we will show that φ_n grows arbitrarily large as n grows. This establishes that the spiral wraps about itself infinitely often.

Comment. φ_{18} is the first angle larger than one full turn. Thus one can stack together a total of 17 triangles before the spiral overlaps. Is this why Theodorus stopped his analysis of irrational roots at $\sqrt{17}$? (See the commentary of newsletter 23.)

From

$$\arctan x = x - \frac{1}{3}x^3 + \frac{1}{5}x^5 - \frac{1}{7}x^7 + \cdots$$

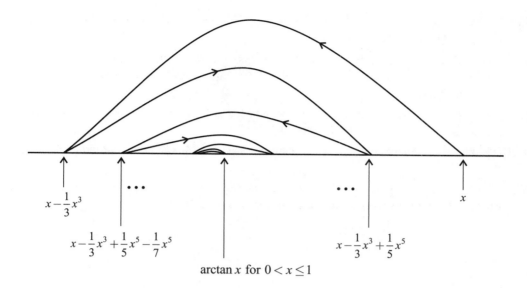

$$x - \frac{1}{3}x^3$$

$$x - \frac{1}{3}x^3 + \frac{1}{5}x^5 - \frac{1}{7}x^5$$

$$x - \frac{1}{3}x^3 + \frac{1}{5}x^5$$

$$x$$

$$\arctan x \text{ for } 0 < x \le 1$$

the diagram makes it clear that $\arctan x \ge x - \frac{1}{3}x^3$ and so if $x = \frac{1}{\sqrt{n}}$ we have

$$\arctan\left(\frac{1}{\sqrt{n}}\right) \ge \frac{1}{\sqrt{n}} - \frac{1}{3n\sqrt{n}} = \frac{1}{\sqrt{n}}\left(1 - \frac{1}{3n}\right) > \frac{1}{2\sqrt{n}}.$$

Thus

$$\varphi_n > \frac{1}{2}\left(1 + \frac{1}{\sqrt{2}} + \cdots + \frac{1}{\sqrt{n}}\right).$$

As $\sum_{k=1}^{\infty} \frac{1}{\sqrt{k}}$ diverges, we have $\varphi_n \to \infty$ as $n \to \infty$, as claimed.

Challenge. Prove that no two turning angles φ_n and φ_m in the spiral differ by a multiple of 2π. (That is, prove that no two hypotenuses in the spiral will coincide.)

References

[BLATNER] Blatner, D., The Joy of π. URL: http://www.joyofpi.com/pilinks.html

[DAVIS] Davis, P.J., *Spirals from Theodorus to Chaos*, A K Peters, Wellesley, MA, 1993.

[GRONAU] Gronau, D., The spiral of Theodorus, *American Mathematical Monthly*, **111**, No. 3 (2004), 230–237.

[HAHN and SCHOENBERGER] Hahn, H. and Schoenberger, K., The ordered distributions of natural numbers on the square root spiral, arXiv:0712.2184v1[math.GM], arXiv.org, 2007.

[HAYASHI] Hayashi, K., Fibonacci numbers and the arctangent function, *Mathematics Magazine*, **76**, No. 3 (2003), 214–215.

[KALMAN and MENA] Kalman, D. and Mena, R., The Fibonacci Numbers – Exposed, *Mathematics Magazine*, **76** (2003), 167–181.

[KNOTT] Knott, R., Pi and the Fibonacci Numbers. URL: http://www.maths.surrey.ac.uk/hosted-sites/R.Knott/Fibonacci/fibpi.html

[KOWALSKI] Kowalski, E., On the "reducibility" of arctangents of integers, *American Mathematical Monthly* **111**, No. 4 (2004), 351–354.

[MACTUTOR] O'Connor, J., Robertson, E. creators: The MacTutor History of Mathematics Archive, University of St. Andrews, Scotland.URL: http://www-history.mcs.st-andrews.ac.uk/BiogIndex.html

[ROY] Roy, R., The discovery of the series formula for pi by Leibniz, Gregory and Nilakantha, *Mathematics Magazine* **63** (1990), 291–306.

[TODD] Todd, J., A problem on arc tangent relations, *American Mathematical Monthly*, **56** (1949), 517–528.

[YOUNG] Young, R., *Excursions in Calculus: An Interplay of the Continuous and the Discrete*, The Mathematical Association of America, Washington D.C., 1992.

2

Benford's Law

PUZZLER

Here is one of my favorite mathematical mysteries.

Consider the powers of two:

1, 2, 4, 8, 16, 32, 64, 128, 256, 512, 1024, 2048, 4096, 8192, 16384, . . .

Do any of them begin with a 7?

If so, which is the first power of two that does? What is the second, and the third? Are there ten powers of two that begin with a seven? Are there infinitely many?

If, on the other hand, no power of two begins with a seven, why not?

ANOTHER PUZZLER

The powers of three begin

1, 3, 9, 27, 81, 243, 729, 2187, 6561, . . .

Several of them end in 1. Do any powers of three end in 01? In 001? In 0001?

TIDBIT: Benford's Law

In 1881, while looking at books of numerical values in his local library – logarithm tables and books of scientific data, for instance – the American astronomer Simon Newcomb observed that the first few pages of tables seemed to be more worn from use than later pages. He surmised that data values that begin with the digit 1 were looked at more often than values beginning with other digits. He thought that this was curious.

The same phenomenon was later independently noticed in 1938 by the physicist Frank Benford, who went further and examined large collections of data tables from a wide variety of sources: population growth data, financial data, scientific observation, and the like. He observed that, with some consistency, about 30.1% of the entries in each data set began with a 1, about 17.6% with a 2, about 12.5% with a 3, all the way down to about 4.6% with a 9. This observation has since become known as Benford's Law.

Filing Taxes. The IRS today uses Benford's law to scan for possibly falsified tax records: About 1/3 of the figures appearing on a tax form should begin with a "1," about 1/6 with a "2", and so forth.

Explaining Benford's Law. Benford's Law can be justified, with relative ease in fact, as long as one is comfortable with logarithms and some sneaky thinking!

Let's work with the powers of two, 1, 2, 4, 8, 16, . . . and show that about 12.5% of them begin with a 3. (We can show that about 30.1% of them begin with a 1 in the same way.)

A power of two begins with a 3 if it lies between 30 and 40, between 300 and 400, between 3000 and 4000, and so on. That is, a power of two, 2^n, starts with a 3 if there is a value k for which

$$3 \times 10^k < 2^n < 4 \times 10^k$$

Taking logarithms we obtain

$$\log 3 + k < n \log 2 < \log 4 + k$$

so

$$k + 0.477 < n \log 2 < k + .602.$$

That is, on a number line we are seeking multiples of $\log 2$ that lie between the positions 0.477 and 0.602 after some whole number k units along the line.

Let's take the number line and curl it into a circle so that all the whole number positions lie on top of one another and all the intervals of interest coincide:

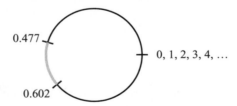

We seek the proportion of the numbers $\log 2$, $2 \log 2$, $3 \log 2$, $4 \log 2$, . . . that lie in the indicated part of the circle.

The circle has circumference one unit and the highlighted portion represents $0.602 - 0.477 = 0.125 = 12.5\%$ of it. If we can show that the multiples of $\log 2$ evenly fill up the circle, then we can legitimately say that 12.5% of the powers of two begin with a 3.

We will show that this is the case.

Now $\log 2$ is quite a complicated number: It is irrational. [If $\log 2 = \frac{a}{b}$, then $10^a = 2^b$, which says that a power of two is divisible by five, which is not true.]

This means that no two multiples of $\log 2$ will ever land on the same point of the circle. [If they do, then we'd have an equation of the form $a + m \log 2 = b + n \log 2$ giving $\log 2 = (b - a)/(m - n)$ a fraction, which is impossible.]

So all the multiples of $\log 2$ – infinitely many of them – mapped on this circle march around the circle and never land in the same position twice.

We're almost done. We just(!) need to show that they spread out around the circle and fill it up without gaps.

To see this... Plot, say, the first 101 multiples of log 2 on the circle. They all can't be 0.01 units or more apart (there are 101 points), so two multiples of log 2 are within a distance of 0.01 of each other. In fact, according to my calculator, 5 log 2 and 98 log 2 are less than 0.01 units apart. This means that if I were to add another 93 multiples of log 2 to consider 191 log 2, and then another 93 to obtain 284 log 2, and so on, I'd obtain a string of multiples of log 2 all less than 0.01 units apart from each other, spreading themselves evenly around the entire circle. The choice 0.01 is arbitrary, and we can find a sequence of multiples of log 2 that spread themselves across the entire circle in steps less than 0.001 units, or 0.00000001 units, and so forth.

So there we have it! The multiples of log 2 fill up the circle in an evenly distributed way and so $\log 4 - \log 3 = 12.5\%$ of them correspond to powers of two that begin with three.

By the same method, $\log 8 - \log 7 = 5.8\%$ of the infinitely many powers of two begin with 7 (answering the opening puzzler), $\log 78 - \log 77 \approx 0.56\%$ of the powers of two begin with 77 and $\log 778 - \log 777 \approx 0.056\%$ with 777. There are also infinitely many powers of two that begin with your birth date and infinitely many that begin with the first billion digits of pi!

Any physical phenomenon that involves powers of a quantity (population growth – powers of e, banking and interest – powers of e, the Fibonacci numbers and their appearance in nature – essentially powers of phi) will follow Benford's law.

Research Corner. How many powers of two have second digit 7? Third digit 7?

COMMENTARY, SOLUTIONS and THOUGHTS

We must hunt for some time to find a power of two that begins with a seven. The first such is 2^{46}, and then it is a surprise to see that 2^{56}, 2^{66}, 2^{76}, 2^{86}, 2^{96} all begin with seven but 2^{106} does not! Next come 2^{149}, 2^{159}, 2^{169}, 2^{179}, 2^{189} but not 2^{199}. Why these runs of ten in the exponents?

They come from the fact that $2^{10} = 1024$ is approximately equal to 1000, and multiplication by 1000 does not change the first digit of a number. Thus if 2^n begins with a 7, then there is a chance that 2^{n+10} will as well. But of course, the error of "+24" in this analysis will eventually build to break this pattern and put an end to any run of tens.

We can see cycles of ten if we list the first digits of the powers of two in a table. Here are the first digits of 2^0 up to 2^{99} listed in rows of ten

```
1  2  4  8  1  3  6  1  2  5
1  2  4  8  1  3  6  1  2  5
1  2  4  8  1  3  6  1  2  5
1  2  4  8  1  3  6  1  2  5
1  2  4  8  1  3  7  1  2  5
1  2  4  9  1  3  7  1  2  5
1  2  4  9  1  3  7  1  3  5
1  2  4  9  1  3  7  1  3  6
1  2  4  9  1  3  7  1  3  6
1  2  5  9  2  4  7  1  3  6
```

Question. What structure lies in a table like this? When a change in a column occurs, is the change sure to be an increase of just one? What about tables of the first digits of powers of three, or powers of 173? (The proof of Benford's Law in the newsletter is not restricted to the powers of two.)

The second puzzler in the newsletter asks about final digits of the powers of three:

Do any of the numbers 1, 3, 9, 27, 81, 243, 729, 2187, 6561, . . . end in 001?

Absolutely!

Divide each of 1, 3, 9, 27, 81, 243, . . . by 1000 and look at their remainders. As there are infinitely many powers of three and only a finite number of possible remainders, there must be two powers, say 3^n and 3^m with $n > m$, that have the same remainder on division by 1000. This means that their difference is a multiple of this number

$$3^n - 3^m = 1000k$$

for some k. Thus

$$3^m \left(3^{n-m} - 1\right) = 1000k.$$

Since 1000 has no factors in common with 3^m, this means that $3^{n-m} - 1$ is a multiple of 1000

$$3^{n-m} - 1 = 1000a$$

for some a Thus $3^{n-m} = 1000a + 1$ is a power of three that ends in 001.

We can repeat this argument for any of the strings 1, 01, 0001, 00001, and so on.

Question. No power of three ends with 002. (Why?) Is there a power of three that ends with 007?

The Exponential Nature of the Fibonacci Numbers

The newsletter claimed that the sequence of Fibonacci numbers, 1, 1, 2, 3, 5, 8, 13, 21, 34, 55, 89, . . . , are exponential in nature and also follow Benford's Law. This requires some explanation.

Let F_n denote the nth Fibonacci number with $F_0 = 1$, $F_1 = 1$, $F_2 = 2$, and so on. We have

$$F_{n+2} = F_{n+1} + F_n$$

We will find a formula for the nth Fibonacci number and so reveal its exponential character.

A standard technique in analyzing a recursive sequence is to ask if there is a geometric sequence $1, x, x^2, x^3, x^4, \ldots$ that satisfies the recursion relation. In this case, we seek a value x that satisfies $x^{n+2} = x^{n+1} + x^n$ for all n. Dividing by x^n we see that this is equivalent to satisfying

$$x^2 = x + 1$$

which occurs for

$$\varphi = \frac{1 + \sqrt{5}}{2} \quad \text{and} \quad \tau = \frac{1 - \sqrt{5}}{2}$$

(The first value is the famous golden ratio [LIVIO].) The two sequences:

$$1, \varphi, \varphi^2, \varphi^3, \ldots$$

$$1, \tau, \tau^2, \tau^3, \ldots$$

follow the same Fibonacci relation: any term (beyond the second) is the sum of the previous two. It is not hard to check that the same is true for any linear combination of the sequences. That is, if a and b are constants and we set

$$G_n = a\varphi^n + b\tau^n$$

then we also have:

$$G_{n+2} = G_{n+1} + G_n$$

for all $n \geq 2$.

Let's now choose constants a and b to create a sequence that has the same start as the Fibonacci numbers. That is, let's choose constants that make $G_0 = 1$ and $G_1 = 1$. If

$$G_0 = a + b = 1$$

$$G_1 = a\varphi + b\tau = 1$$

algebra gives $a = \frac{\varphi}{\sqrt{5}}$ and $b = -\frac{\tau}{\sqrt{5}}$.

We now have a sequence G_n that starts with the same initial values as the Fibonacci sequence and follows the same recursive relation as the Fibonacci sequence. It must be the Fibonacci sequence!

$$F_n = \frac{\varphi}{\sqrt{5}} \cdot \varphi^n - \frac{\tau}{\sqrt{5}} \cdot \tau^n = \frac{\varphi^{n+1} - \tau^{n+1}}{\sqrt{5}} = \frac{1}{\sqrt{5}}\left(\left(\frac{1+\sqrt{5}}{2}\right)^{n+1} - \left(\frac{1-\sqrt{5}}{2}\right)^{n+1}\right).$$

This is Binet's formula for the Fibonacci numbers. (This formula is today attributed to the French mathematician Jacques Binet (1786–1856) though it was known by other scholars decades earlier ([LIVIO]).)

Because $|\tau| = \left|\frac{1-\sqrt{5}}{2}\right| \approx 0.618 < 1$, $\tau^{n+1} \to 0$ as $n \to \infty$. Thus for n large we have

$$F_n \approx \frac{\varphi^{n+1}}{\sqrt{5}}.$$

Challenge. Prove that for each n, F_n is the nearest integer to $\frac{\varphi^{n+1}}{\sqrt{5}}$.

Thus the Fibonacci numbers are essentially exponential.

Comment. For more on Benford's law see [WEISSTEIN1]. The uniformity of the distribution of multiples of log 2 about the circle allows us to assign meaningful percentages (see [WEISSTEIN2]) and the proof of this uniformity was only alluded to in the newsletter. Full details can be found on page 390 of [HARDY and WRIGHT].

Final Thought

I once gave the following challenge as an e-mail puzzle for the St. Mark's Institute followers.

LIKING MY THREES:

I like numbers that contain the digit 3. The number $N = 751153$ contains the digit three, as does its double $2N = 1502306$, $3N = 2253459$, $4N = 3004612$, and $5N = 3755765$. (Alas, $6N$ does not contain a three!)

a) Find a number N such that N, $2N$, $3N$, $4N$, $5N$ and $6N$ each contain the digit three.

b) Find a number N such that N, $2N$, $\ldots$, $10N$ each contain the digit three.

c) For each k is there an N with the property that the multiples N, $2N$, $3N$, $\ldots$, kN contain the digit three?

d) Is there an N such that all integer multiples of N contain the digit 3?

We can check that for $N = 19507893$ each of N, $2N$, $3N$, $\ldots$, $25N$ contain a three, which answers parts a) and b).

Challenge. What is the smallest number that answers a) and the smallest number that answers b)?

It is fairly easy to settle part d) in the negative:

Lemma 1. *There is no positive integer N so that all its multiples, N, $2N$, $3N$, $\ldots$, contain the digit three.*

Proof. Given a positive integer N we'll construct a multiple of N that does not contain the digit three.

Divide each of $1, 11, 111, 1111, 11111, \ldots$ by N and look at their remainders. As there are only a finite number of possible remainders, two of the numbers in the infinite list have the same remainder, so their difference is a multiple of N.

But this difference is of the form $11 \ldots 100 \ldots 0$ and so is a multiple of N that contains no three.   □

To answer part c) we need an intermediate result.

Lemma 2. *For each positive integer k there is an integer N so that kN contains the digit three.*

Proof. Look at the infinite list of numbers $3, 33, 333, 3333, 33333, \ldots$. Divide them by k and look at their remainders. As there are only a finite number of possible remainders, two numbers in the list leave the same remainder. Their difference, a number of the form $33 \ldots 300 \ldots 0$, is a multiple of k and contains a three.   □

For example

For $k = 1$, $N = 3$ has $1 \cdot N$ containing a three.

For $k = 2$, $N = 15$ has $2N$ containing a three.

For $k = 3$, $N = 10$ has $3N$ containing a three.

For $k = 4$, $N = 75$ has $4N$ containing a three.

We can construct a number that works simultaneously for the four values of k by stringing together the values of N we have so far, separating them by zeros. For example, for

$$N = 75 \ \ 00000 \ \ 10 \ \ 00000 \ \ 15 \ \ 00000 \ \ 3$$

we have that each of N, $2N$, $3N$ and $4N$ contains the digit three. With the aid of Lemma 2 we can construct a single value N whose multiples up to any value we choose will contain the digit three. (We need to place sufficiently many zeros between the terms in the construction so as to absorb any carries that might occur. We can construct a pattern for the number of zeros that will suffice.)

This establishes

Theorem. *For each positive integer k there is an integer N such that each of the multiples $N, 2N, 3N, \ldots, kN$ contain the digit three.*

References

[HARDY and WRIGHT] Hardy, G. H., and Wright, E. M., *An Introduction to the Theory of Numbers*, Clarendon Press, Oxford, England, 1979.

[LIVIO] Livio, M., *The Golden Ratio: The Story of Phi, the World's Most Astonishing Number,* Broadway Books, New York, NY, 2002.

[WEISSTEIN1] Weisstein, E. W. Benford's Law from *MathWorld*-A Wolfram Web Resource. URL: http://mathworld.wolfram.com/BenfordsLaw.html

[WEISSTEIN2] Weisstein, E. W. Equidistributed sequences from MathWorld-A Wolfram Web Resource. URL: http://mathworld.wolfram.com/EquidistribuedSequence.html

3
Braids

PUZZLER: A Weird Language

The language of ABABA uses only two letters, A and B, and any combination of them is a word. (Thus, for instance, ABBBBABAA and BBB are both words.) Also, strangely, a blank space, , is considered a word.

The language has the property that any word that ends in ABA or in BAB has the same meaning as the word with them deleted. (Thus, for example, BBABA and BB are synonyms.) Also, any two consecutive As or Bs can be deleted from a word without changing its meaning. (Consequently, BAABBBA, BAABA, BBA, and A are equivalent words.)

Warm-up. Show that BA and A are synonyms.

Challenge. How many words of distinct meaning does this bizarre language possess?

[This is a strange puzzle. Its relevance is made clear in the next section.]

TIDBIT: Try This!

Take three strings, two colored red and one yellow, and tie them to the back of a chair so that the yellow strand lies in the middle position. Braid the three strands in any manner you care to choose. That is, cross adjacent strands over or under each other in any organized or disorganized fashion. Make sure when you are done that the yellow strand is in the middle position. Tie the three ends to a wooden spoon.

Here's something amazing!

Fact. *No matter what braid you create (with the middle stand ending in the middle position), it is possible to untangle it by maneuvering the spoon back and forth between the strands.* (Try it!)

This shows that the braid produced with free ends could have been produced with the ends fixed in place!

Explanation. (It is good to have three strings and a wooden spoon in hand as you read this section.) We can read a braid from top down as a sequence of crossings. Let L denote the crossing of a strand in the left-most position over or under the strand in the middle position, and R the crossing of the right two strands.

LLRLR

It does not matter if an L is an under or an over crossing. Experimentation shows that we can convert an under crossing to an over crossing, and vice versa, by pushing the wooden spoon between the right two strands just below the crossing. (Try it!) The same is true for an R crossing. Thus any braid is encoded in a well-defined manner as a string of Ls and Rs.

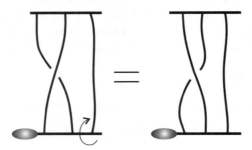

If a braid possesses two L crossings in a row, then the spoon-pushing trick shows that we can convert one of the Ls to an under-crossing and the other an over-crossing. Clearly, two such consecutive Ls is equivalent to a braid with those two Ls omitted.

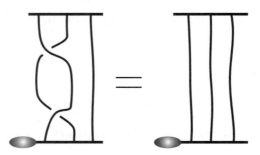

Similarly, we can remove any two consecutive Rs from a braid.

If a braid ends in the sequence LRL, then a 180° rotation of the spoon deletes it.

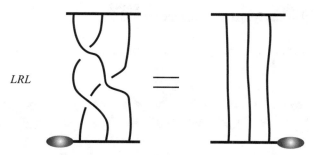

LRL

Similarly we can remove the string RLR from the end of a braid.

So ... A code for a braid is a string of letters L and R. Since we can delete two neighboring Ls or Rs, a braid is equivalent to one that alternates L and R. Since we can delete RLR or LRL from the end of a braid, this leaves only five possibilities for a braid: LR, RL, R, L, or the blank space, the untangled state.

We can show that the braid LR is equivalent to R, and RL to L. Thus any braid reduces to either R, L, or the untangled state. Neither of the first two options have yellow strand returning to the middle position. Thus, the only permissible option for a braid with the middle strand ending in the middle position is for the braid to be equivalent to the untangled state!

Students from the St. Mark's Institute research group went further and proved:

> *Given a string of Ls and Rs for the code of a braid change every second entry of the code to an R if it was an L, and to an L if it was an R. (For example, the code LRLLRLRLL becomes LLLLRRRRRL.)*

> *Count the number of Ls in the modified code and subtract from it the number of Rs. If the difference is a multiple of three, then the original braid can be untangled. If it is one more than a multiple of three, then the braid is equivalent to a single L, and equivalent to R if it is one less.*

As far as I am aware, this is a new technique for quickly analyzing three-braids.

Research Corner

1. Up a Count

Analyze which braids with four strands can be completely untangled if their ends are tied to a spoon.

2. Flat Braids

Take a rectangular piece of felt cut in it two slits. Make the braid shown with no free ends. Notice that the strands of this braid remain flat, that is, no strand possesses an internal twist.

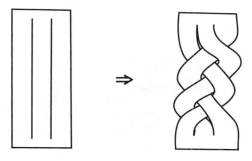

Classify all flat braids that can be made with three strands.

Comment. The high-school research group made some progress with this tough question.

COMMENTARY, SOLUTIONS and THOUGHTS

Flat Braids

At every place where I work, as a college professor or as a high-school teacher, I hang from my office door or on my classroom walls a collection of felt strips braided as shown:

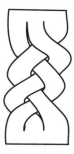

Each day, after leaving my office for a spell, I would find those strips unbraided! A physical model of this flat braid is an irresistible lure. (It is worth the effort to make a large model out of poster paper to put on public display.)

The mathematical analysis of flat braids seems to be tricky. Students of the St. Mark's research group found a way to encode twists in braid strands and add to the L/R coding scheme described in the newsletter. Although they were able to establish some good results about the algebra of these codes, a full analysis was too complex. (Might you find a better approach to studying flat braids?) The students came to suspect that the flat braid shown above is the only example that can exist, up to repetition of the pattern along the braid. (But is it?)

String Braids

Braids of strings with no free ends, as described in the newsletter, are discussed in [SHEPPERD]. I also discuss aspects of this material in [TANTON].

Here is a variation of the three-strings problem:

Three strings attached to the ceiling of the room are braided in some haphazard manner and taped to a wooden spoon. When is it possible to untangle them by holding the spoon fixed in space and maneuvering the strings about it?

We have two fundamental maneuvers: lift an end string over the wooden spoon (either in a forwards or backwards direction) or do the same for the middle string:

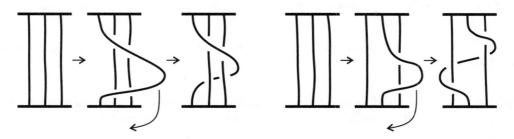

Each maneuver introduces four crossings in the strands. There are also two maneuvers that we can perform within a braid:

Let's give a crossing a value of $+1$ if, for the two strands involved, the left strand passes over the right strand, and value -1 in the other case. Define the *index* of a braid as the sum of the values of its crossings. Each braid maneuver we've described changes the index of a braid by ± 4, or not at all. Thus only those braids whose index is a multiple of four have hope of being untangled. (Of course, in addition, the ends of the strings must be tied to the spoon in the correct order for the hope to continue!)

This result rules out hope that one particular type of braid can be untangled:

> *Hang a spoon from a ceiling with three untangled strands and give it one full turn of rotation. It is impossible to untangle this braid the braid that results by holding the spoon fixed in space and maneuvering the strings about it.*

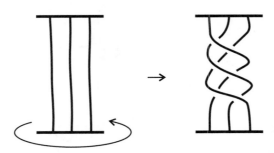

The braid has index six.

Here is a surprise:

Dirac's String Trick. *It is possible to untangle the braid that results from rotating the spoon through TWO full rotations.*

This works for any number of strings tied to the spoon. (See [NEWMAN], [TANTON], and also [BOLKER].)

Exploration.   For three strings hanging from the ceiling with the spoon held fixed in space, denote a crossing with the left two strands L if it has value $+1$ and L^{-1} if it has value -1, and denote a crossing of the right two strands R and R^{-1} similarly. Show that $L^{-1}R^{-1} = LR$ and $R^{-1}L^{-1} = RL$. What other relations between the symbols can we write down?

References

[BOLKER] Bolker, E. The spinor spanner, *The American Mathematical Monthly*, **80**, No. 9. (1973), 977–984.

[NEWMAN] Newman, M. H.A., "On a string problem of Dirac," *The Journal of the London Mathematics Society*, **17**, Part 3, No. 67 (1942), 328–333.

[SHEPPERD] Shepperd, J.A.H., "Braids which can be plaited with their threads tied together at each end," *Royal Society of London Proceedings*, *A*, **265** (1962), 229–244.

[TANTON] Tanton, J., *Solve This: Math Activities for Students and Clubs*, Mathematical Association of America, Washington D.C., (2001).

Clip Theory

PUZZLER: Believing in Patterns

This is one of my favorite teasers. I use it in all of my extracurricular courses for teachers and students. Certainly patterns must be true!

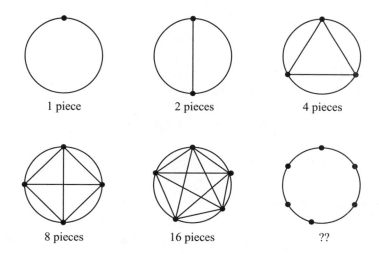

| 1 piece | 2 pieces | 4 pieces |
| 8 pieces | 16 pieces | ?? |

Here we are assuming that the dots are spaced so that the maximal number of pieces appears when each pair of dots is connected by a straight line.

Going Further. Seven dots? Eight dots? Nine dots?

Bending the Rules. This month's puzzler is a surprise. (Really do try it!) A greater surprise still lies hidden in the mathematics if one is willing to be flexible in one's play, so flexible as to flex the lines!

Here's a diagram of three curved lines ($L = 3$) connecting pairs of points in a circle.

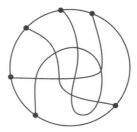

This circle is divided into 12 pieces ($P = 12$). There are 8 points of intersection within the circle ($I = 8$) and in this picture

$$P = L + I + 1.$$

The same is true for each of the pictures in the puzzler (for example, in the fifth picture $P = 16$, $L = 10$ and $I = 5$) and for <u>most</u> any picture you draw. (Try it!)

A complication lies with multiple intersections. In

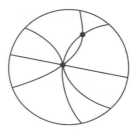

$L = 4$ and $P = 9$ and it seems $I = 2$, invalidating our claim. But are the two intersections of equal weight? One of the two is a single road atop another whereas the other is three roads atop another. If we set $I = 1 + 3 = 4$ then the equation $P = L + I + 1$ again holds true!

Conjecture. *In any diagram of curved or straight lines drawn across a circle, $P = L + I + 1$, as long as intersections are counted with appropriate weight.*

Actually, the conjecture is a theorem!

Proof. The equation $P = L + I + 1$ holds for a diagram with no lines ($P = 1$, $L = 0$, and $I = 0$). Let's see if it remains valid as we draw new lines, adding one line at a time. In starting a new path two things occur as soon as it intersects itself or intersects a pre-existing path: the count of intersections increases by one and the count of pieces increases by one. The formula $P = L + I + 1$ remains balanced. The equation remains valid when the line returns to the circumference of the circle: the count of lines increases by one, balanced by the fact that a final region is split in two. The formula $P = L + I + 1$ holds all the way through the process of drawing any new line.

So this does it! The equation $P = L + I + 1$ starts out valid and never changes as lines are added. It remains true then for all diagrams.   □

Getting a Formula. If you completed the opening puzzle you may have discovered the sequence

1, 2, 4, 8, 16, 31, 57, 99, 163, . . .

for the number of pieces from n dots drawn on a circle with straight lines. Is there a formula for these piece numbers?

We have $P = L + I + 1$, so the challenge lies in finding formulas for L and I. In the straight line case, each line determines and is determined by a pair of boundary dots:

The formula for the number of ways to select two objects from n is

$$L = \frac{n!}{2!\,(n-2)!} = \frac{n\,(n-1)}{2}$$

Each intersection point determines and is determined by a choice of four boundary points. (Think about this.)

The formula for selecting four objects from n is

$$I = \frac{n!}{4!\,(n-4)!} = \frac{n\,(n-1)\,(n-2)\,(n-3)}{24}.$$

Thus the sequence of piece numbers is given by

$$P = L + I + 1 = \frac{n\,(n-1)}{2} + \frac{n\,(n-1)\,(n-2)\,(n-3)}{24} + 1,$$

a friendly looking formula!

Bonus. What is the sum of the first five entries of each row of Pascal's triangle?

1									= 1
1	1								= 2
1	2	1							= 4
1	3	3	1						= 8
1	4	6	4	1					= 16
1	5	10	10	5	1				= 31
1	6	15	20	15	6	1			= 57
1	7	21	35	35	21	7	1		= 99
1	8	28	56	70	56	28	7	1	= 163

Coincidence?

Research Corner. Draw a self-intersecting loop of any kind.

Discover – and prove – interesting things about the count of pieces and the count of intersections. Two intertwined loops?

COMMENTARY, SOLUTIONS and THOUGHTS

When a group of young students (ages 10–13) explored the circle on the dots problem with me and discovered for themselves its generalization to curved lines, they not only counted intersection points (I), lines (L) and pieces (P), but also the number of circles (C), which was always one! They called their study CLIP theory as they had proved

$$C + L + I = P.$$

The newsletter shows that $L = \binom{n}{2}$ and $I = \binom{n}{4}$. We can write $C = \binom{n}{0}$. As every entry in Pascal's triangle is the sum of the two entries immediately above it, we have

$$P = \binom{n-1}{0} + \binom{n-1}{1} + \binom{n-1}{2} + \binom{n-1}{3} + \binom{n-1}{4}$$

(with the understanding that $\binom{n}{k}$ is zero if $k > n$). This explains why each count is the sum of the first five entries of a row of the triangle.

   If one draws a self-intersecting loop the count of finite regions produced (pieces, P) is always one more than the count I of intersection points (counted with multiplicity): $P = I + 1$. We can establish this in much the same way as we proved CLIP theory.

Simple Loops

Let's call a self-intersecting loop *simple* if each point of intersection is passed through only twice as we trace the curve (no multiple crossings) and no two sections of the curve touch each other tangentially. The simple self-intersecting loops have an interesting property:

> *In tracing the curve starting at any intersection point P, we pass through an <u>even</u> number of intersection points before returning to P.*

Challenge. Establish this claim.

Warning: The result need not be true for simple loops drawn on non-flat surfaces. For example, we can draw a loop on the surface of a torus (donut) for which this fails. (Try it!) Thus this claim must involve a deep fundamental property of planar curves. Which one? For help, see [TANTON, Chapter 29].

Number the points of intersection of a simple self-intersecting loop 1, 2, 3, ... Starting at a non-intersection point P trace the curve and list the intersection points we encounter until we return to P. Since the curve is simple, each intersection point is listed exactly twice and, by the property stated above, each label appears once in an odd position on the list and once in an even position.

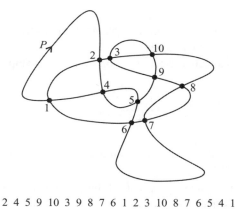

$$2\ 4\ 5\ 9\ 10\ 3\ 9\ 8\ 7\ 6\ 1\ 2\ 3\ 10\ 8\ 7\ 6\ 5\ 4\ 1$$

$$= \frac{2\ \ 5\ \ 10\ \ 9\ \ 7\ \ 1\ \ 3\ \ \ 8\ 6\ 4}{4\ \ 9\ \ 3\ \ \ 8\ \ 6\ \ 2\ \ 10\ \ 7\ 5\ 1}$$

Thus if we split the list of intersection labels into two, the odd and the even positioned elements, the second list is a permutation of the first.

The following diagrams show that every permutation of one, two and three elements occurs.

Order 1   **Order 2**

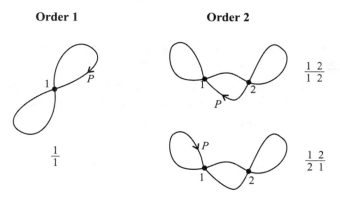

$$\frac{1}{1}$$

$$\frac{1\ 2}{1\ 2}$$

$$\frac{1\ 2}{2\ 1}$$

Order 3

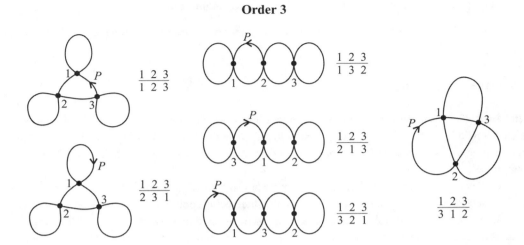

Challenge. Show that every permutation on four elements arises from some loop walk. Only five distinct shapes appear among the 24 examples.

Not every permutation of elements arises as a loop walk. For example, there is no loop with the permutation

$$\frac{1\ 2\ 3\ 4\ 5}{3\ 4\ 5\ 1\ 2}.$$

Challenge. There is only one other permutation on five elements that cannot be realized as a loop walk. Which one?

Of the $6! = 720$ permutations on six elements all but 36 can be realized as loop walks (and 44 distinct loop shapes are used), and of the $7! = 5040$ permutations on seven elements all but 571 can be so realized (using a total of 180 distinct loop shapes). (See [PEGG] and [ROSENSTIEHL and TARJAN].)

Challenge. Can the missing permutations be realized as walks along loops on the surface of a torus?

References

[PEGG] Pegg Jr., E., Contributor, Gauss Code Loops from The Wolfram Demonstrations Project. URL: http://demonstrations.wolfram.com/GaussCodeLoops/

[ROSENSTIEHL and TARJAN] Rosenstiehl, P. and Tarjan, R.E., Gauss codes, planar Hamiltonian graphs, and stack-sortable permutations, *Journal of Algorithms* **5** (1984), 375–390.

[TANTON] Tanton, J., *Solve This: Math Activities for Students and Clubs,* Mathematical Association of America, Washington D.C., 2001.

5
Dots and Dashes

PUZZLER

1. The sequence of square numbers begins 1, 4, 9, 16, 25, ... and the n th square number is n^2.

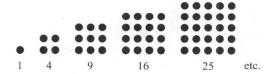

The sequence of non-square numbers begins 2, 3, 5, 6, 7, 8, 10, 11, 12, 13, 14, 15, 17, ...

a)   What's the 100th non-square number?

b)   Find a formula for the n th non-square number

2. The sequence of triangular numbers begins 1, 3, 6, 10, 15, ... and the n th triangular number is $\frac{1}{2}n(n + 1)$.

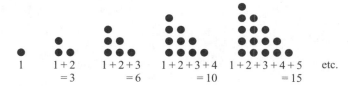

The sequence of non-triangular numbers begins 2, 4, 5, 7, 8, 9, 11, ...

a)   What's the 100th non-triangular number?

b)   Find a formula for the n th non-triangular number.

TIDBIT

In the list of prime numbers

$$[p_n] : 2, 3, 5, 7, 11, 13, 17, ...$$

(here p_n means the n th prime) let q_n denote the number of primes less than n. For example, $q_{10} = 4$ because there are 4 primes less than ten. This gives the frequency sequence for the primes

$$[q_n]: 0, 0, 1, 2, 2, 3, 3, 4, 4, 4, 4, 5, \ldots.$$

Now let's compute the frequency sequence of the frequency sequence! There are 2 numbers among $[q_n]$ less than one, 3 numbers less than two, 5 numbers less than three, 7 numbers less than four, 11 numbers less than five, and so on. If you keep checking you see that the sequence of primes reappears!

$$[p_n]: 2, 3, 5, 7, 11, 13, 17, \ldots$$

More is true! Take the sequence $[p_n]$ and its frequency sequence $[q_n]$ and add 1 to the first element of each, 2 to the second element of each, 3 to the third, 4 to the fourth, and so on. We obtain the sequences

$$[P_n]: 3, 5, 8, 11, 16, 19, 24, 27, 32, \ldots$$

$$[Q_n]: 1, 2, 4, 6, 7, 9, 10, 12, 13, 14, \ldots$$

(So $P_n = p_n + n$ and $Q_n = q_n + n$.)

ALL the counting numbers $1, 2, 3, 4, \ldots$ appear, split among these two sequences with no repeats! Mathematicians call these "complementary sequences."

The lovely phenomenon is not unique to the list of prime numbers.

Theorem. *Write down a sequence of non-negative whole numbers that never decreases. (You may repeat entries multiple times.) Call the sequence $[p_n]$.*

i) *The frequency sequence of the frequency sequence of $[p_n]$ is $[p_n]$.*

ii) *Adding position numbers to the entries of $[p_n]$ and to the entries of its frequency sequence produces complementary sequences.*

Let's try this on the following sequence (which I just made up!):

$$[p_n]: 1, 2, 2, 2, 3, 3, 6, 7, 7, 7, 7, 9, 11, 11, 14, 14, 14, 14, 14, 14, 14, 14, 15, \ldots.$$

Its frequency sequence is:

$$[q_n]: 0, 1, 4, 6, 6, 6, 7, 11, 11, 12, 12, 14, 14, 14, 22, \ldots$$

(There are 0 entries in $[p_n]$ less than one, 1 entry less than two, 4 entries less than three, and so on.)

Now compute the frequency sequence for $[q_n]$ to see that is $[p_n]$. (Do this!)

Add the position numbers $1, 2, 3, 4, 5, \ldots$ to each of the sequences $[p_n]$ and $[q_n]$. This gives

$$[P_n]: 2, 4, 5, 6, 8, 9, 13, 15, 16, 17, 18, 21, 24, 25, 29, 30, 31, 32, 33, 34, 35, 36, 38, \ldots$$

$$[Q_n]: 1, 3, 7, 10, 11, 12, 14, 19, 20, 22, 23, 26, 27, 28, 37, \ldots$$

Yep. Complementary!

Explanation. In 2005, three St. Mark's students, Charles Zodda, Eric Rudyak, and Jae Shin You, came up with the following innovative proof: A sequence $[p_n]$ can be encoded as a string of dots and dashes. For example, the sequence 1, 2, 2, 2, 3, 3, 6, 7, 7, 7, 7, 9, 11, 11, 14, 14, 14, 14, 14, 14, 14, 14, 15, ... is encoded as:

$$*|*|||*||***|*||||**|**||***||||||||*| \cdots$$

Here the first dash has 1 dot to its left, the second dash has 2 dots to its left, the third dash has 2 dots to its left, the fourth 2 dots, the fifth 3 dots, the sixth 7 dots, and so on. (Thus the n th dash is placed so that it has p_n dots to its left.)

In the diagram

p_n = *number of dots to the left of the nth dash.*

An entry q_n of the frequency sequence is the number of entries of $[p_n]$ with value less than n. In the diagram, this is the number of dashes with fewer than n dots to their left. If you think about it, this means:

q_n = *the number of dashes to the left of the nth dot.*

This is the definition of p_n with "dot" and "dash" interchanged. Thus "frequency" means "interchange symbols." So the dot-dash diagram for the frequency sequence is the original diagram with dots and dashes switched. (Try this! Draw the dot-dash diagram for 0, 1, 4, 6, 6, 6, 7, 11, 11, . . .) Doing this twice (taking the frequency of the frequency) returns us to the original diagram and hence to the original sequence!

Further. The n th dash has $n - 1$ dashes to its left and, by definition, p_n dots to its left. Thus the n th dash is in position $(n - 1) + p_n + 1 = p_n + n$. Interchanging "dots" and "dashes" tells us, by the same reasoning, that the nth dot lies in position $q_n + n$. Thus, if we list the counting numbers along the top of a dot-dash diagram, the locations of the dashes give us the values of $P_n = p_n + n$ and the locations of the dots the values $Q_n = q_n + n$. The list of counting numbers is split among two sequences given by the positions of the dots and the dashes!

```
1 2 3 4 5 6 7 8 9 10 11 12 13 14
                                    ...
* | * | | | * | |  *   *   *  |  *
```

Research. Can anything of interest be deduced from diagrams composed of strings of three symbols? How about two-dimensional arrays of symbols?

COMMENTARY, SOLUTIONS and THOUGHTS

The dots and dashes method allows us to find formulas for the non-square and non-triangular numbers.

Let $[P_n]$ be the sequence of square numbers and $[Q_n]$ be its complementary sequence, the non-squares:

$$[P_n] : 1, 4, 9, 16, \ldots$$

$$[Q_n] : 2, 3, 5, 6, 7, 8, 10, \ldots$$

We can think of these sequences as arising from a diagram of dots and dashes:

```
1  2  3  4  5  6  7  8  9  10  11  12  13  14
*  |  |  *  |  |  |  |  *  |   |   |   |   |
```
...

For each n, set $p_n = P_n - n = n^2 - n$ and $q_n = Q_n - n$. Given the work in the newsletter we have arranged matters so that $[q_n]$ is the frequency sequence of $[p_n]$.

$q_n = $ the number of entries of $[p_n]$ with value less than n

$= $ the number of values k so that $p_k < n$

$= $ the number of values k so that $k^2 - k < n$.

Since we are speaking only of integers, $k^2 - k < n$ also means that $k^2 - k + \frac{1}{4} < n$. Thus:

$q_n = $ the number of values k so that $k^2 - k + \frac{1}{4} < n$

$= $ the number of values k so that $\left(k - \frac{1}{2}\right)^2 < n$

$= $ the number of values of k so that $k < \sqrt{n} + \frac{1}{2}$

$$= \left\lfloor \sqrt{n} + \frac{1}{2} \right\rfloor$$

where $\lfloor x \rfloor$ is the largest integer no larger than x. If we take $\langle x \rangle$ to mean the integer nearest to x, then a little thought shows that

$$\left\lfloor x + \frac{1}{2} \right\rfloor = \langle x \rangle.$$

Thus we have a formula for q_n,

$$q_n = \langle \sqrt{n} \rangle,$$

and hence also for Q_n, the nth non-square number

$$Q_n = q_n + n = \langle \sqrt{n} \rangle + n = \langle \sqrt{n} + n \rangle.$$

For example, the 100th non-square number is $\langle 10 + 100 \rangle = 110$.

Challenge. Derive the formula for the n th non-square number by a method that doesn't use results about frequency sequence. (St. Mark's Institute followers found a number of derivations.)

For the non-triangular numbers... Let $[P_n]$ be the sequence of triangular numbers and $[Q_n]$ its complementary sequence of non-triangulars

$$[P_n] : 1, 3, 6, 10, 15, \ldots$$

$$[Q_n] : 2, 4, 5, 7, 8, \ldots.$$

We know $P_n = \frac{1}{2}n(n + 1)$. We want a formula for Q_n.

Set $p_n = P_n - n = \frac{1}{2}n^2 - \frac{1}{2}n$ and $q_n = Q_n - n$. We know that $[q_n]$ is the frequency sequence of $[p_n]$. Thus

$$q_n = \text{the number of values } k \text{ so that } \frac{1}{2}k^2 - \frac{1}{2}k < n$$

$$= \text{the number of values } k \text{ so that } k^2 - k < 2n.$$

The argument is now the same, except we are working with $2n$ instead of n. This gives the nth non-triangular number as

$$Q_n = \left\langle \sqrt{2n} + n \right\rangle.$$

For example, the 100th non-triangular number is $\langle \sqrt{200} + 100 \rangle = 114$.

Challenge. If $\langle \sqrt{n} + n \rangle$ gives the non-square numbers and $\langle \sqrt{2n} + n \rangle$ gives the non-triangular numbers, what non-numbers are given by $\langle \sqrt{3n} + n \rangle$?

Challenge. What's the nth non-cube? What's the nth non-kth power (for $k \geq 4$)?

Beatty Sequences

In 1926 Samuel Beatty observed that complementary sequences of positive integers occur in another context ([BEATTY]).

Let α be a positive irrational number greater than one and set β so that $\frac{1}{\alpha} + \frac{1}{\beta} = 1$. (Algebra shows that β is also an irrational number greater than one.) Set $P_n = \lfloor n\alpha \rfloor$ and $Q_n = \lfloor n\beta \rfloor$. Then

The sequences $[P_n]$ and $[Q_n]$ are complementary.

For example, with $\alpha = \sqrt{2}$ and $\beta = 2 + \sqrt{2}$ we obtain

$$[P_n] : 1, 2, 4, 5, 7, 8, 9, 11, 12, 14, 15, 16, 18, 19, \ldots,$$

$$[Q_n] : 3, 6, 10, 13, 17, 20, \ldots$$

Establishing the claim is not difficult.

Since α is greater than one, no integer is repeated within the sequence $[P_n]$. Similarly, no integer is repeated among the values of $[Q_n]$. But could an integer appear in both sequences?

Suppose there are integers n and m so that $\lfloor n\alpha \rfloor = \lfloor m\beta \rfloor$. Call the common value k. We then have

$$k \leq n\alpha < k + 1,$$

$$k \leq m\beta < k + 1.$$

(The inequalities are strict since α and β are each irrational.) So

$$\frac{k}{\alpha} < n < \frac{k+1}{\alpha},$$

$$\frac{k}{\beta} < m < \frac{k+1}{\beta}.$$

Adding and using $\frac{1}{\alpha} + \frac{1}{\beta} = 1$ yields the absurdity

$$k < n + m < k + 1.$$

Thus no integer appears more than once among the two sequences.

Could a positive integer be missing? Suppose integer a is skipped over by both $[P_n]$ and $[Q_n]$. This means that there are integers n and m with

$$n\alpha < a \quad \text{and} \quad (n + 1)\alpha \geq a + 1,$$

$$m\beta < a \quad \text{and} \quad (m + 1)\beta \geq a + 1.$$

So

$$n < \frac{a}{\alpha} \quad \text{and} \quad \frac{a + 1}{\alpha} < n + 1,$$

$$m < \frac{a}{\beta} \quad \text{and} \quad \frac{a + 1}{\beta} < m + 1$$

and adding yields

$$n + m < a \quad \text{and} \quad a + 1 < n + m + 2,$$

from which we deduce that a is an integer strictly between $n + m$ and $n + m + 1$. This absurdity shows that no positive integer can be missed by the sequences.

Challenge. Let γ be a positive irrational number (not necessarily greater than one). Find a formula for the frequency sequence of $\lfloor n\gamma \rfloor$. (Actually ... need γ be irrational?)

Challenge. In the newsletter we assumed that no value is repeated infinitely often in $[p_n]$. Would there be a problem if the sequence were eventually constant?

Reference

[BEATTY] Beatty, S., Problem 3173, *American Mathematical Monthly*, **33** (3), (1926), 159.

6

Factor Trees

PUZZLER: Factor Trees

In grade school students draw factor trees. Here is a tree for 36,000:

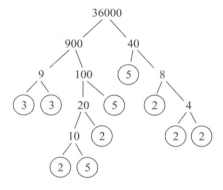

At each stage we split the number at hand into a pair of factors, halting at the primes. (This forces the tree to stop. Good thing that 1 is <u>not</u> considered prime!) The tree allows us to write the starting number as a product of primes. Here we see $36000 = 2 \cdot 2 \cdot 2 \cdot 2 \cdot 2 \cdot 3 \cdot 3 \cdot 5 \cdot 5 \cdot 5$. What is astounding (though most people don't seem to think so) is that despite possible different choices we can make along the way, this produces the same list of primes. (Thus, for example, if we started the factor tree with $36000 = 360 \times 100$ we'd still, allegedly, obtain precisely five 2s, two 3s and three 5s.) It is not at all obvious that matters should work out this way! (See the next page for indication of what can go wrong.)

a) Draw a different factor tree for 36000 (do it—you'll need it for the rest of this puzzler) and verify that the same primes appear.

Factor trees hold other surprising invariants.

b) The factor tree I drew contains nine pairs of numbers:

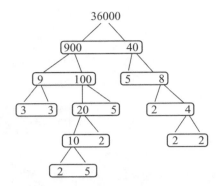

I bet the factor tree you drew for a) also contains nine pairs. Why must all factor trees for a number contain the same number of pairs?

b)   In the example above, there are four pairs with both numbers even. I bet the same is true for the factor tree you drew. Why must each factor tree have the same number of even/even pairs?

c)   In my tree there are two pairs with both numbers divisible by 5. How do I know that the same must be true for your tree?

d)   The product of the numbers in a pair is the number just above the pair. (900 × 40, for example, is 36000 and 9 × 100 is 900.) Let's multiply one less than each number in a pair and sum the results. In my tree this gives

$$899 \times 39 = 35061$$
$$8 \times 99 = 792 \qquad 1 \times 3 = 3$$
$$4 \times 7 = 28 \qquad 9 \times 1 = 9$$
$$2 \times 2 = 4 \qquad 1 \times 1 = 1$$
$$19 \times 4 = 76 \qquad 1 \times 4 = 4$$
$$\text{SUM} = 35{,}978$$

Do the same for your tree and verify that you obtain the same sum. Why must this be so?

On Uniqueness of Factor Trees

We are led to believe in grade school that all factor trees decompose a number into the same set of primes, no matter the choices we make along the way. This is not at all obvious. Suppose Lulu and Gary agree to devote ten hours to factoring 4542847263897240000002300100000000.

Is it transparent that they must obtain exactly the same list of primes in the end? Here's a cute example to illustrate my point:

In the country of Evenastan only even numbers exist! If you ask a citizen of that land to count to ten, she will respond: 2, 4, 6, 8, 10. (And if you ask her to count to 11, she'll only give you a puzzled look. There is no such thing as "11" in Evenastan.)

In this world of evens, some numbers factor and some don't. For example, 24 factors (4 × 6) but 26 does not. (Remember, "13" does not exist.) Those numbers that factor are called e-composite (short for "Evenastan-composite") and those that don't, e-prime.

Exercise. List the first twenty e-primes.

Just as in the U.S., young children in Evenastan are taught to draw factor trees. Here is a factor tree for 40:

This shows that 40 factors into e-primes:

$$40 = 2 \times 2 \times 10.$$

Unlike children in the U.S., young Evenastan scholars realize that factor trees are <u>not</u> unique.

Here are two factor trees for the number 400 showing that it decomposes into e-primes in at least two different ways.

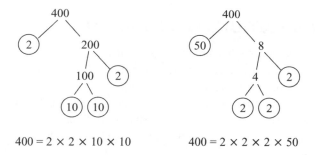

$$400 = 2 \times 2 \times 10 \times 10 \qquad 400 = 2 \times 2 \times 2 \times 50$$

Challenge. *What is the smallest number in Evenastan that factors into e-primes in more than one way?*

So... If factor trees are not unique in Evanastan, what makes us think they are unique in our world of evens and odds? Maybe the examples we examine in school are too small to encounter a problem?

Greek mathematician Euclid (ca. 300 B.C.E.) pondered questions of both geometry and number theory and realized that there is something significant and deep to prove about our counting numbers. Although the details are beyond the space of this newsletter, Euclid established that prime factorizations from factor trees are unique for our arithmetic. What we were implicitly told to assume true is actually true!

It is very easy to train our students—and ourselves!—to be passive in mathematics learning, not to query accepted norms and not think to question mathematics. Even if one cannot prove a claim right away, it is okay to hold a question at the back of one's mind to ponder upon over time. Mysteries can be joyous. This is what makes math an intensely human experience.

Challenge. Are factor trees in Oddvanistan unique?

Research Corner

Suppose factor trees are based on triples rather than pairs, stopping at primes and at products of two primes (semiprimes!). For example

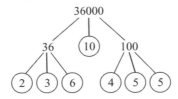

These trees certainly are not unique in our arithmetic. Any invariants nonetheless?

COMMENTARY, SOLUTIONS and THOUGHTS

When working with middle- and high-school teachers it always seems to come as a surprise that I question the uniqueness of factor tree results. The example of Evenastan is effective in justifying why a query is worthwhile, in fact, necessary!

The e-primes in Evenastan are 2, 6, 10, 14, 18, 22, 26, 30, 34, 38, 42, 46, 50, 54, 58, 62, 66, 70, 74, 78..., and 36 is the smallest number that factors two different ways: $36 = 6 \times 6 = 2 \times 18$. Anyone playing with very small examples in Evenastan (numbers smaller than 36) might think that all factor trees yield unique results.

So perhaps, in our system of all counting numbers, the examples we've seen are too small. How do we know that there isn't some large number after which the uniqueness of factorization breaks down?

As mentioned in the newsletter, Euclid established the fundamental (and remarkable!) result that our counting numbers are sufficiently tight to provide unique prime factorizations. Any beginning course in number theory goes through the details. (See also [TANTON1], or the classic text [HARDY and WRIGHT].)

Since the primes in Oddvanistan are primes in our arithmetic, our odd primes, Euclid's result applies in this world as well: Oddvanistanian factor trees are unique.

Invariants in Factor Trees

I based this section of a newsletter on the article [TANTON2], converting the additive results there to multiplicative results of factor trees. In my experience, students enjoy the Pile Splitting games in that article. Older students enjoy following the ideas and applying them to other systems, such as factor trees.

Here is a general means for developing invariant results.

Let A_1, A_2, A_3, ... be a sequence of numbers. For each split in a factor tree

(here $N = ab$), write the quantity

$$A_a + A_b - A_N.$$

If we do this for each pair in a factor tree and sum the results, we see that the answer is independent of the choices made along the way: all intermediate nodes, which correspond to our choices, cancel. For example, with the tree

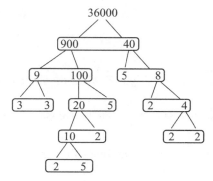

we obtain the sum

$$(A_{900} + A_{40} - A_{36000}) + (A_9 + A_{100} - A_{900}) + (A_5 + A_8 - A_{40})$$
$$+ (A_3 + A_3 - A_9) + (A_{20} + A_5 - A_{100}) + (A_2 + A_4 - A_8)$$
$$+ (A_{10} + A_2 - A_{20}) + (A_2 + A_2 - A_4) + (A_2 + A_5 - A_{10})$$

which is

$$A_2 + A_2 + A_2 + A_2 + A_2 + A_3 + A_3 + A_5 + A_5 + A_5 - A_{36000}.$$

These are the terms corresponding to the prime factors of 36000 and the number 36000 itself. <u>All</u> factor trees for 36000 give this result!

Now it is just a matter of choosing interesting sequences to play with.

1. Work with the sequence 1, 1, 1, 1, ... (that is, set $A_a = 1$ for all a).

Then the quantity $A_a + A_b - A_{ab}$ always equals 1 and so by writing down these quantities and summing, we are counting the number of pairs in a tree. The result is $A_2 + A_2 + A_2 + A_2 + A_2 + A_3 + A_3 + A_5 + A_5 + A_5 - A_{36000} = 9$ for all 36000 trees.

2. Work with the sequence $A_a = \begin{cases} 0 & \text{if } a \text{ is even} \\ 1 & \text{if } a \text{ is odd} \end{cases}$.

Then $A_a + A_b - A_{ab}$ is 0 if both a and b are even, and is 1 otherwise. The sum we create counts pairs that contain at least one odd number. The final result is $A_2 + A_2 + A_2 + A_2 + A_2 + A_3 + A_3 + A_5 + A_5 + A_5 - A_{36000} = 5$. As the total number of pairs is 9, this means that there are 4 even/even pairs in all trees.

3.   Work with the sequence $A_a = \begin{cases} 0 & \text{if } a \text{ is a multiple of 5} \\ 1 & \text{otherwise} \end{cases}$.

It shows that there are two pairs with both numbers divisible by 5.

Comment. There are other ways to establish the results. For example, the prime factorization of 36000 contains five 2s and three 5s and these primes must be pulled out one at a time.

4.   Work with the sequence $A_a = 1 - a$.

Then $A_a + A_b - A_{ab}$ equals $(a - 1)(b - 1)$. The sum of these strange products is invariant!

References

[HARDY and WRIGHT] Hardy, G.H. and Wright, E. M., *An Introduction to the Theory of Numbers*, 5th ed., Clarendon Press, Oxford, 1979.

[TANTON1] Tanton, J., *THINKING MATHEMATICS! Volume 1:Arithmetic = Gateway to All*, www.lulu.com, 2009.

[TANTON2] Tanton, J., A dozen questions about pile splitting, *Math Horizons*, September 2004, 28–31.

7

Folding Fractions and Conics

PUZZLER: Trisecting an Angle

Sally draws a straight line from the bottom left corner of a blank piece of paper. She challenges Terrell to make a crease in the paper that bisects the angle formed (that is, cuts that angle exactly in half). "Easy" says Terrell as he lifts up the bottom edge, aligns it with the straight mark, and folds.

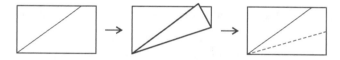

"Now," chortles Terrell in response, "I challenge you to make *two* crease marks that divide your angle exactly into thirds."

It can be done using nothing more than creases in the paper. How?

TIDBIT: Trisecting a Strip

Without the aid of any tools it is very difficult (impossible?) to fold a strip of paper into exact thirds. But here's a technique that allows you to make a crease as close to one third as you please. (So close that no one could ever detect the error—even with a micrometer!).

1. *Guess where the 1/3 mark lies along the strip of paper and record the guess with a crease.*
2. *Pick up the right end of the strip, align it with your guess, and fold to produce a new crease halfway between your guess and the right end.*
3. *Pick up the left end of the strip, align it with this second crease, and fold. This gives a third crease closer to the true 1/3 position than your guess. It has one quarter the error!*
4. *Repeat the sequence of right and left folds to reduce the error as much as desired.*

To explain why this works, call the left end of the strip position 0 and the right end position 1 and suppose the initial guess lies at $1/3 + E$. (Here E represents an error.) The first right fold produces a crease at a position $\frac{2}{3} + \frac{E}{2}$, half way between $1/3 + E$ and 1. It is close to the two-third mark with half the error. A fold from the left then produces a crease at position: $\frac{1}{3} + \frac{E}{4}$, and the error has been slashed by a factor of four. A sequence of just RL folds reduces an error by a factor of more than a million!

ANOTHER PUZZLER: Dyadic Fraction Folding

A fraction whose denominator is a power of two is called *dyadic*. They can be obtained by summing a finite number of fractions from the list

$$\frac{1}{2}, \frac{1}{4}, \frac{1}{8}, \frac{1}{16}, \ldots$$

For example, $\frac{13}{16} = \frac{1}{2} + \frac{1}{4} + \frac{1}{16}$ is dyadic, as is $\frac{1}{128}$, but the fraction $\frac{1}{3}$ is not.

Dyadic fractions have finite decimal expansions in binary arithmetic. For instance, $13/16 = .1101$ in base two, and $.101011$ represents the dyadic fraction:

$$\frac{1}{2} + \frac{1}{8} + \frac{1}{32} + \frac{1}{64} = \frac{43}{64}.$$

Here's something amazing:

Take a strip of paper one foot long. To create a crease at 13/16 read its base two representation $13/16 = .1101$ backwards as a set of instructions! Interpret 1 as "fold right" and 0 as a "fold left." Look!

- **1**: Lifting the **right** end of the strip and folding produces a mark at position $\frac{1}{2}$.
- **0** : Lifting the **left** end and folding to the previous crease gives a mark at position $\frac{1}{4}$.
- **1**: Lifting the **right** end and folding to the previous crease produces a mark at position 5/8.
- **1**: Lifting the **right** and folding to the previous crease produces a mark at position 13/16, as desired.

This always works! Create a fold at position 43/64 by following the instructions .101011, and a fold at position 1/128 by the instructions .0000001.

Challenge 1. Why does this work?

Challenge 2. What is the binary representation for the fraction 1/3? How does its expansion relate to the repeated cycle of instructions "RL" given in today's tidbit?

Folding Conics

1. Wanna see a parabola? Take a blank sheet of paper, draw a straight line and draw a dot not on the line. Fold the paper to position that dot somewhere on the line. (You might have to hold the paper up to the light.) Make a crease and unfold the paper. Repeat at least twenty times, positioning the dot at different locations along the line. The creases you make outline a perfect parabola!

Comment. It is a easier to use the bottom edge of the paper as the chosen line. Keep folding it up to the dot at different angles.

2. Wanna see a hyperbola? On another sheet draw a circle and a dot outside it. Fold the dot onto different positions around the circle, making creases as you go. The two branches of a hyperbola emerge!

3. Wanna see an ellipse? On a third sheet draw a circle and a dot *inside* the circle (preferably not the center). Fold the dot to different positions around the circle making creases as you go. What are the two foci of the ellipse you see?

Challenge. Why do these folding exercises work?

Research Corner

If we draw two dots on a page and then fold the page so as to align the dots, the crease produced represents all points on the page equidistant from those two dots. That is, we can think of a line as a set of points equidistant from two dots.

A parabola is defined in a similar manner: it is the set of points equidistant from a line and a dot. (What might we mean by "distance to a line"?) People usually call the dot the *focus* and the line the *directrix*.

What curves result if instead we look for the set of points equidistant from a dot and a circle? (How should we define "distance to a circle"?)

Draw two objects on a page, say, a dot and a square, a circle and an ellipse, or two circles of different radii. What curves emerge as you investigate the set of all points equidistant from your two objects?

What can be said about equidistance for three objects?

COMMENTARY, SOLUTIONS and THOUGHTS

Here's how to trisect an angle with origami ([HULL1]). As in the newsletter we assume that the angle is defined by a line L drawn from bottom left corner A of a piece of paper and the bottom edge of that paper.

Fold a crease parallel to the bottom edge and fold a second parallel crease at half its height. (Both are easy to do.) Call the distance between the two creases a and the point on the left edge at height $2a$ point B. Take the bottom left corner of the sheet and fold so that B lies on line L (call the location of this point C) and A lies on the crease at height a to define a point D. The line connecting A to D precisely trisects the original angle. Bisecting angle DAC completes the trisection.

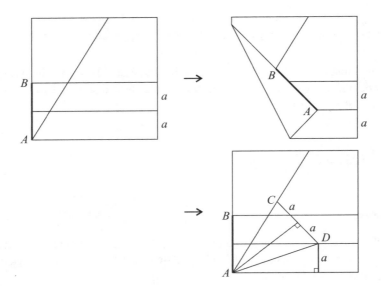

To prove that this works it suffices to establish that the three right triangles shown in the third diagram are congruent.

Let E be the location of the crease mark on the bottom edge created by the fold that takes $\overline{BA}$ to $\overline{CD}$. Triangle AED is isosceles. Label its congruent base angles w.

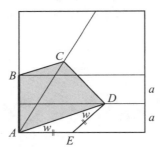

We have $\overline{BA} \cong \overline{CD}$ and $m\angle BAD = m\angle CDA = 90 - w$. Thus $ABCD$ is an isosceles trapezoid and so has congruent diagonals: $AC = BD$. By Pythagoras's theorem $BD = AD$ and so triangle CAD is isosceles. Thus any two of the three right triangles we are examining share two common lengths and so are congruent.

For more ways to trisect an angle check out Jim Loy's webpage [LOY].

Dyadic Fraction Folding

Given a strip of paper one foot long, our two operations, fold left and fold right, have the following effects when starting with a crease at position x along the strip:

Fold left produces a new crease at position $\frac{x}{2}$.

Fold right produces a new crease at position $x + \frac{1-x}{2} = \frac{x}{2} + \frac{1}{2}$.

In terms of binary representations, halving a number shifts all terms in its expansion one place to the right. For example, the fraction $\frac{5}{8} = \frac{1}{2} + \frac{1}{8}$ with binary representation .101, when halved, becomes $\frac{5}{16} = \frac{1}{4} + \frac{1}{16}$ with representation .0101. Rephrasing this:

A fold left has the effect of inserting a zero to the right of the decimal point of the binary representation of the crease and shifting the remaining terms one place to the right.

A right fold halves a number and adds the fraction $\frac{1}{2}$, which in binary is 0.1. Thus

A fold right has the effect of inserting a one to the right of the decimal point and shifting all remaining terms one place to the right.

Starting with the fraction 0 (a clean strip of paper with a hypothetical crease at the left end) we can create any dyadic fraction we choose following these maneuvers. For example, to create $\frac{13}{16} = .1101$, start with 0 and apply the following folds.

Fold Right:	0	$\rightarrow$	.1
Fold Left:	.1	$\rightarrow$	.01
Fold Right:	.01	$\rightarrow$	.101
Fold Right:	.101	$\rightarrow$	.1101

This is equivalent to looking at the representation .1011 and reading it backwards interpreting 1 as fold right and 0 as fold left!

The Fraction One-third

In base two, 1/3 has non-terminating representation .0101010101.... (See newsletter 24.) From this we can obtain dyadic fractions that approximate one third to any degree of accuracy. For example

$$\frac{1}{3} \approx .01 \qquad \text{(Fold Right then Left)},$$

$$\frac{1}{3} \approx .0101 \qquad \text{(Fold Right, Left, Right, Left)},$$

$$\frac{1}{3} \approx .010101 \quad \text{(Fold Right, Left, Right, Left, Right, Left)}.$$

This shows that the repeated cycle of right and left folds converges to 1/3.

Comment. It does not matter how we start the procedure. If we start with the fraction $0.abcd\ldots$ and apply the sequence of right/left folds, it too will converge to the fraction at position 1/3:

$$0.abcd\ldots \rightarrow 0.01abcd\ldots \rightarrow 0.0101abcd\ldots \rightarrow 0.010101abcd\ldots \rightarrow \cdots$$

We can use the technique to approximate other non-dyadic fractions. For example, the fraction 5/7 has binary representation .101 101 101 101 ... showing that a sequence of RLR folds will produce creases that converge to 5/7, and 1/15 has representation .000100010001 ... so repetitions of the sequence RLLL converge to it.

Question. What does the infinite repetition of a right fold R suggest about 0.1111111111...?

Comment. As hinted, folding dyadic fractions make a reappearance in newsletter 24. They have a surprising connection to water pouring problems.

Folding Conics

A *parabola* is defined as the set of all points equidistant from a given point F (the focus) and a given line L (the directrix). The distance of a point from a line is measured as the length of a line segment perpendicular to the line.

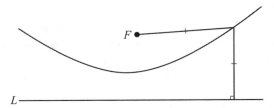

Suppose we are given a point F and a line L marked on a piece of paper and we fold F onto point P on the line to make a crease, the perpendicular bisector of $\overline{FP}$.

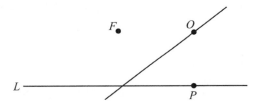

There is just one point O on the crease that is the same distance from F as it is from L, the point O for which $\overline{OP}$ is perpendicular to L. This means the the crease has precisely one point in common with the parabola defined by F and L. That is, the crease is a tangent line to the parabola. Folding F to different locations along the line produces an envelope of tangent lines that reveal the curve.

An *ellipse* with foci F and G is defined as the set of all points P whose distances PF and PG sum to a fixed value. A *hyperbola* with the same foci is the set of all points Q whose distances QF and QG have a fixed positive difference. (Two branches arise by considering each of $QF - QG$ and $QG - QF$ as being the positive value.)

There is an alternative way to define an ellipse:

An ellipse is the set of points equidistant from a circle and a point F inside it.

The distance from an interior point to a circle is the shortest distance from it to the circle.

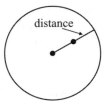

distance

To see this:

> Suppose point F is inside a circle with center G and P is a point equidistant from F and the circle. Then $PF + PG$ is a constant, the radius of the circle.

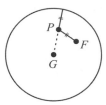

Let's now examine the folding activity described in the newsletter.

> *To see an ellipse ... On a blank sheet draw a circle and a dot inside the circle (preferably not the center). Fold the dot to different positions around the circle making creases as you go. The shape of an ellipse appears.*

As for the parabola, the creases form an envelope of tangent lines to the ellipse. To understand why, consider a point F inside a circle with center G. Suppose we have folded F to R on the circumference of the circle and formed a crease, the perpendicular bisector of $\overline{FR}$.

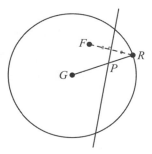

We show that there is precisely one point on this crease-line that is equidistant from F and the circle. Let P be the point of intersection of the radius $\overline{GR}$ and the crease. Since $PF = PR$ it is equidistant as hoped. Suppose there is a second point Q on the crease-line equidistant from F and the circle. Let S be the point on the circle so that $\overline{GS}$ is a radius through Q.

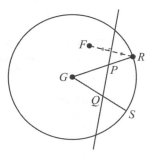

Equidistance gives $QS = QF$. Since Q is on the crease line we also have $QF = QR$. Thus triangles SQR and SGR are both isosceles, but that is impossible!

Thus the crease line is a tangent line to the ellipse.

The branches of a hyperbola can also be defined in terms of equidistance from a point F to a circle. Take F as a point outside the circle. For the present define the distance of a point as the positive difference between the radius of the circle and the distance of the point from the center, the distance between the closest point on the circle to the given point.

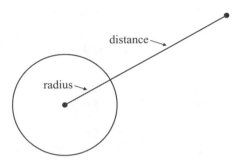

If P is equidistant from F and the circle, then we see that the difference $PG - PF$ is the radius of the circle. We have one branch of the hyperbola.

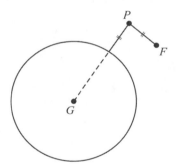

Challenge. Suppose we measured the distance of a point P from a circle as the *largest* distance between P and some point on the circle. Show that the set of points P equidistant from F and the circle gives the second branch of the hyperbola: the set of points P with $PF - PG$ equal to the fixed radius of the circle.

Challenge. Suppose we use the alternative measure of distance from a circle for a point F inside the circle. What changes when we attempt to define an ellipse using equidistance this way?

The newsletter describes a third folding activity:

To see a hyperbola . . . On a blank sheet draw a circle and a dot outside the circle. Fold the dot to different positions around the circle making creases as you go. The shape of a hyperbola appears.

Challenge. Complete the necessary work to justify this.

The reflection properties of conics can also be explored by folding. See newsletter 19, "On Reflection" for this or [TANTON, Chapter 29] for details. (Also see [HULL2].)

Challenge. Two circles are drawn on a page. What curve appears as the set of points P equidistant from them?

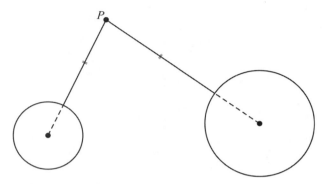

What if the circles overlap or if one circle sits inside the other?

Challenge. Three circles are drawn on a page. Will there be a point equidistant from them? Investigate when it is possible to draw a circle mutually tangent to three circles.

References

[LOY] Loy, J., Trisection of an angle. URL: http://www.jimloy.com/geometry/trisect.htm

[HULL1] Hull, T., A note on "impossible" paper folding, American Mathematical Monthly, **103** (1996), 242–243.

[HULL2] Hull, T., *Project Origami: Activities for Exploring Mathematics*, A.K. Peters, Newton, MA, 2006.

[TANTON] Tanton, J., *GEOMETRY: Volume II,* www.lulu.com, 2010.

Folding Patterns and Dragons

PUZZLER: Paper Folding Sequence

Take a strip of paper and imagine its left end is taped to the ground. If we pick up the right end, fold the strip in half, and unfold it, the paper has a valley crease in its center. Label it 1.

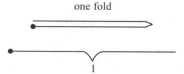

one fold

Suppose we perform two folds, lifting the right end up over to the left end. When we unfold the paper, we find three creases: two valley creases and one mountain crease. Label the mountain crease as 0.

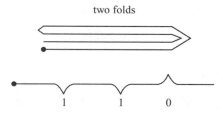

two folds

For three folds we obtain

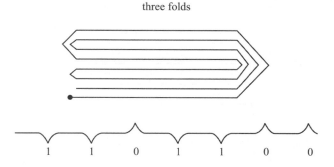

three folds

The sequence for four folds turns out to be 110110011100100.

a)   What is the sequence of 1s and 0s for five folds?

b)   How many digits are in the one-hundred fold sequence? How many of those digits are 1s? How many are 0s?

c)   Look at the sequences we have for one, two, three, and four folds:

$$1$$
$$1\ 1\ 0$$
$$\mathbf{110}\ 1\ 100$$
$$\mathbf{1101100}\ 1\ 1100100$$

Each begins with the entire previous sequence. Coincidence?

Also ... The second portion of each sequence is the dual of the first portion: it is the first portion written backwards with the zeros changed to ones and ones changed to zeros. Is this always so?

Comment. If this is true, then to write the six-fold sequence, write down a five-fold sequence, add a 1, and then write it backwards switching 1s and 0s.

d)   Look at the four-fold sequence

$$110110011100100$$

The first term is 1, the third term 0, the fifth term is 1, the seventh term is 0, and so on. That is, the odd terms give the alternating sequence 1,0,1,0,1,0, 1,0. The even terms are 1,1,0,1,1,0,0 which appear to be the previous fold sequence!

Prove that each fold sequence is the intertwining of the alternating sequence 1, 0, 1, 0, 1, 0, ... with the previous sequence.

Comment. This gives another way to write the next fold sequence from a given one.

e)   What is the 112th digit of the one-hundred fold sequence?

Discussion and Answers

The properties of the folding sequences are astounding. Let's prove them here. (So don't read on if you wish to ponder the puzzler further!)

One fold produces two layers of paper, two folds four layers of paper, and in general, n folds give 2^n layers.

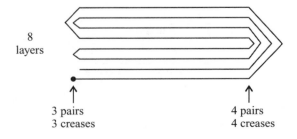

8
layers

3 pairs
3 creases

4 pairs
4 creases

On the right side of a fold layers come in $\frac{1}{2} \cdot 2^n$ pairs to make that many creases and on the left side there is one less crease. The total number of creases after n folds is $\frac{1}{2} \cdot 2^n + \frac{1}{2} \cdot 2^n - 1 = 2^n - 1$, so there are $2^n - 1$ entries in the nth folding sequence.

The left set of creases on a given fold is just the set of creases from the previous fold. The creases on the right are the new creases.

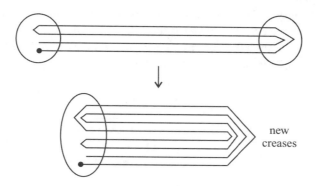

new
creases

Starting at the solid dot in the diagram, if we trace along the line we first pass through a new crease on the right (upwards), then through the first crease of the previous fold, then through another new crease on the right (downwards), then the second crease of the previous fold, and so on. That is, we pass alternately through a new crease and an old crease. Also, we pass through the new creases alternately upwards and downwards (giving alternate valley and mountain creases) and we pass though the old creases in the same order that they were before. This explains the intertwining feature of part d). To justify c), again imagine tracing the line of the paper starting at the solid dot. As we move to the middle crease, we see that we are tracing the outline of the previous fold, and so follow the same creases as the previous fold. (Imagine that two layers of paper are glued together. This reduces the fold to the previous case.)

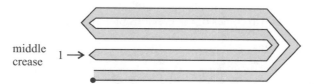

middle
crease $\quad 1 \rightarrow$

Adding colour makes three fold look like two folds

We then meet the middle crease, 1, and then trace the same sequence of creases in reverse order and in reverse direction (thus turning all valleys to mountains and mountains to valleys). Thus

$$\begin{array}{ccccc} \text{new} & = & \text{old} & +1+ & \text{dual of} \\ \text{sequence} & & \text{sequence} & & \text{old sequence} \end{array}$$

All the 1s in the old sequence are 0s in the dual, and vice versa. Thus there is an equal number of 0s and 1s in "old sequence + dual." The 1 in the middle shows that: *In any sequence there is precisely one more 1 than 0s*. Since there $2^n - 1$ creases in all, $2^{n-1} - 1$ are zeros and 2^{n-1} are ones.

To identify particular elements in a sequence use d):

$$\text{new sequence} = \text{old sequence} \quad \text{intertwined with } 1, 0, 1, 0.$$

The 1st, 3rd, 5th, 7th, etc entries of any sequence are 1, 0, 1, 0, etc. and the even entries come from the previous sequence. The 2m-th entry of any sequence is the m-th entry of the previous sequence. So the 112th entry in the 100-fold sequence was the 56th entry in the previous sequence, which was the 28th in the one before, which was the 14th in the one before that, which was the 7th in the one before that, which was a 0. In general: *If we write $m = 2^a b$ with b odd, then the mth entry of a folding sequence is 1 if m is one more than a multiple of four and 0 if one less.*

Research Corner. Examine the sequences that arise from repeated triple folds:

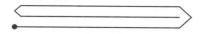

COMMENTARY, SOLUTIONS and THOUGHTS

There is wonderful geometry to be discovered if we tease apart the folds described in the newsletter and use two dimensions. After folding a strip in half multiple times from right to left, open up the strip partway so that each crease corresponds to a turn of 90°.

Marking the left end of the strip with a dot and tracing along the curve, each valley crease corresponds to a left turn of 90° and each mountain crease to a right turn of 90°. The pattern described in part c) of the newsletter is apparent in the geometry that results: each picture in the sequence comes from joining two copies of the previous picture with one rotated 90°.

L	LLR	LLR L LRR	LLRLLRR L LLRRLRR

It is curious that as we repeat this process, the construct never overlaps on itself. That is, no line segment ever lands on top of a previously constructed line segment:

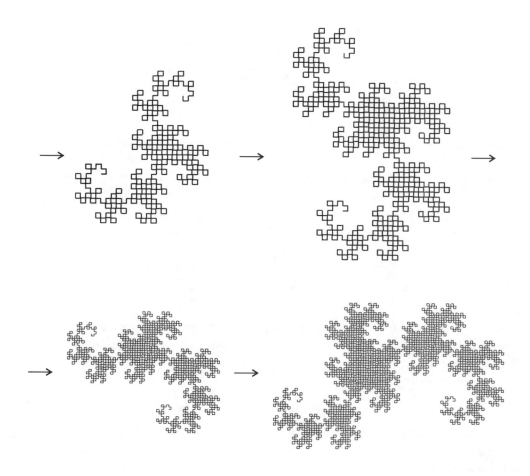

To see why this is so, we use pattern d) described in the newsletter:

To obtain the next sequence from a given one, intertwine it with the alternating sequence.

To conduct this geometrically, replace each line segment in one design with the two legs of a right isosceles triangle:

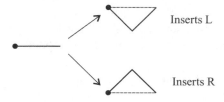

It is clear that no line segment will be traced more than once.

L   L L R   L L R L L R R

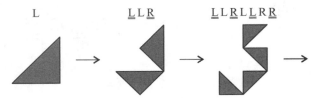

(This construction produces the same designs, but all line segments are reduced in length, by a factor of $\frac{1}{\sqrt{2}}$, at each iteration.)

Something marvelous can be seen if we color each triangle that appears. We obtain a sequence of tiles.

L   L L R   L L R L L R R

Four copies of each tile, at least for the three that we have drawn, can be arranged in a pinwheel fashion about a single point to fit inside a square without overlap (except at the vertices).

This holds for all the tiles that appear in the sequence. To see why, note that in each picture the square is divided into a number of small right triangles and that they have an alternating pattern, shaded and white. Experimentation and thought shows that applying the iteration step to the four tiles within the square divides each small isosceles triangle in two, shading one half, leaving the other half white, and preserving the alternating shading pattern for the triangles throughout the square. No overlap of shaded triangles occurs.

Although the area of any one tile is one-eighth of the area of the square, we can always find a tile for which four copies of it fill up the square to any degree of density we desire. (We can make the white right isosceles triangles, the spaces, as small as we please.)

The paper-folding curve taken to the limit, or sometimes the fractal figure obtained from shaded isosceles triangles, is call Heighway's dragon curve. It is a space-filling curve with the property that four copies of the curve fit together about a single point. Since squares tile the plane, placing four copies of the curve within each square tile shows that the dragon curve also tiles the entire plane. This explains the pictures one sees on internet sites about this curve. (In [RIDDLE], for example.)

More about the mathematics of the curve appears in [DAVIS and KNUTH].

Counting Permutations

I sent the following puzzler out by e-mail to the followers of the St. Mark's Math Institute.

Respect the Inequalities

Here are ten empty spaces with nine inequality signs between them.

a)   *Is it possible to insert the numbers 1 through 10 into the empty spaces and respect the inequality signs? Might there be more than one way to do this?*

b)   *Do questions like these always have solutions? That is, given N empty spaces and N − 1 inequality signs between them will there be at least one arrangement of the numbers 1 through N that satisfies the inequalities?*

c)   *Given one of these puzzles is there a way to determine how many solutions it has?*

d)   *Given 10 empty spaces, what sequence of inequality signs produces the most solutions?*

It is easy to see that puzzles like this have at least one solution:

> Suppose we are to place N distinct values in N spaces about $N − 1$ inequality signs. If the first sign is $>$, place the largest number available in the first slot, otherwise insert the smallest number available if the first sign is $<$. We now have the task of placing $N − 1$ distinct integers among a string of $N − 2$ inequality signs. This is a smaller problem and we follow the same procedure.

For our problem the method produces the solution

$$10 > 1 < 2 < 9 > 8 > 3 < 7 > 4 < 6 > 5.$$

This is not the only solution. In fact there are 23,671 solutions in all!

Comment. There are $N!$ permutations of N symbols and there are 2^{N-1} possible sequences of inequalities for $N − 1$ spaces. Thus the average number of solutions is $\frac{N!}{2^{N-1}}$. If all the inequality signs are identical, the problem possesses exactly one solution, which is the minimum value.

Alex Smith, as a tenth-grader at St. Mark's School, picked up on this challenge and has been considering it for two years now. He developed a simple "tableaux" method for computing the number of permutations that fit a given set of inequalities. (His method is in the literature. See [ATKINSON1], [ATKINSON2], [BRIGHTWELL and WINKLER], and [MILLER, SLOANE and YOUNG], for instance.) It proves that the zig-zag inequalities $< > < > < > \ldots$ possess the largest number of solutions. Most impressive of all, Alex has discovered the following astounding connection to paper folding.

Theorem (A. Smith).

a)   *The average number of solutions to an inequality problem on N symbols, $\frac{N!}{2^{N-1}}$, is an integer if and only if N is a power of two.*

b)   *Take a strip of paper and fold it in half multiple times, either from right to left or from left to right or a mixture of both. Regard each valley crease as a > sign and each mountain crease* *as a $<$ sign. Then the sequence of creases produced represents an inequality problem with the average number of solutions.*

Proof. a) Of the numbers 1 through N, $\lfloor \frac{N}{2} \rfloor$ of them are multiples of two, $\lfloor \frac{N}{4} \rfloor$ of them are multiples of four, and so forth. (Here $\lfloor x \rfloor$ denotes the largest integer no larger than x.) Thus two appears $\lfloor \frac{N}{2} \rfloor + \lfloor \frac{N}{4} \rfloor + \lfloor \frac{N}{8} \rfloor + \cdots$ times in the prime factorization of $N!$. If $N = 2^k$ is a power of two, then the count is $2^{k-1} + 2^{k-2} + \cdots + 2 + 1 = 2^k - 1 = N - 1$ and $\frac{N!}{2^{N-1}}$ is an (odd) integer. If, on the other hand, $N = 2^k m$ for an odd integer m greater than 1, then the count of factors of two in $N!$ is

$$m \left(2^{k-1} + 2^{k-2} + \cdots + 2 + 1 \right) + \left\lfloor \frac{m}{2} \right\rfloor + \left\lfloor \frac{m}{4} \right\rfloor + \left\lfloor \frac{m}{8} \right\rfloor + \cdots$$

$$= N - m + \left\lfloor \frac{m}{2} \right\rfloor + \left\lfloor \frac{m}{4} \right\rfloor + \left\lfloor \frac{m}{8} \right\rfloor + \cdots .$$

Since m is odd $\lfloor \frac{m}{2} \rfloor = \frac{m}{2} - \frac{1}{2}, \lfloor \frac{m}{4} \rfloor \leq \frac{m}{4} - \frac{1}{4}, \lfloor \frac{m}{8} \rfloor \leq \frac{m}{8} - \frac{1}{8}$, and so on. As m cannot be one greater than every power of two, at least one inequality is strict. Thus the count of factors of two in $N!$ is strictly less than

$$N - m + \frac{m}{2} - \frac{1}{2} + \frac{m}{4} - \frac{1}{4} + \frac{m}{8} - \frac{1}{8} + \cdots = N - 1.$$

Consequently $\frac{N!}{2^{N-1}}$ cannot be an integer.

b) It is easy to see that if we reverse the sequence of inequalities the number of solutions does not change. Nor does it change if we flip the sign of every inequality in the string. (Replace the number k from 1 through N with the number $N + 1 - k$.) Thus the dual of a string of inequalities (flip each sign and reverse the order) has the same number of solutions as the original sequence.

Following the notation of the newsletter, a sequence of valleys and creases (1s and 0s) obtained from folding is generated from the iteration rule

$$\begin{array}{ccccccc} \text{new} & & \text{old} & & \text{the symbol 1} & & \text{dual of} \\ \text{sequence} & = & \text{sequence} & + & \text{or} & + & \text{old sequence} \\ & & & & \text{the symbol 0} & & \end{array}$$

where 1 is inserted in the center if the fold is from the right, and 0 if the fold is from the left. We are to read the 1s and 0s as inequalities.

We will show that a sequence of inequalities that arises from this iteration process has precisely the average number of solutions. This is true if there is only one inequality sign (one fold). Suppose the claim is true for any sequence of N symbols arising from folding where N is a power of two. We will show that the problem on $2N$ symbols given by the $2N - 1$ inequalities arising from one more fold has $\frac{(2N)!}{2^{2N-1}}$ solutions.

Matching each sequence with its dual shows that we can match any sequence that arises from folding with central term 1 with a sequence with central term 0, and vice versa. Thus, if S is a sequence of $N - 1$ inequalities that arises from folding and S^* is its dual, then the count of solutions to $S + 1 + S^*$ matches the count of solutions to $S + 0 + S^*$. We know by the induction hypothesis that S (and S^* too) has precisely $\frac{N!}{2^{N-1}}$ solutions.

Imagine the permutation problem has its central inequality changed to $\neq$, $S + $ " $\neq$ " $+ S^*$. To solve it, we must select N symbols from 1 through $2N$ to use in the solution of S, count the number of solutions to S with them, and count the number of solutions to S^* with the remaining symbols.

We see that $S + $ " $\neq$ " $+ S^*$ has

$$\binom{2N}{N} \frac{N!}{2^{N-1}} \cdot \frac{N!}{2^{N-1}}$$

solutions. But the set of solutions to $S + $ " $\neq$ " $+ S^*$ is the union of the set of solutions to $S + 1 + S^*$ and to $S + 0 + S^*$, and these sets have equal size. Thus the set of solutions to a new sequence that arises from S by applying one more fold is

$$\frac{1}{2} \cdot \binom{2N}{N} \frac{N!}{2^{N-1}} \cdot \frac{N!}{2^{N-1}} = \frac{(2N)!}{2^{2N-1}},$$

as hoped! $\square$

The "A. Smith Conjecture". *If a sequence of $N - 1$ inequalities has precisely $\frac{N!}{2^{N-1}}$ solutions, then it arose as a folding pattern.*

This remains unproven. Alex continues to work on it and has discovered a host of interesting auxiliary results about counting solutions, not just on linear strings of inequalities, but also on directed graphs.

Final Comment

For decades there seemed to be a general belief that a strip of paper cannot be folded in half more than eight times. This was proved wrong in 2001 when Britney Gallivan, then a junior in high school, succeeded in folding a 4000 foot length of toilet paper in half twelve times. (See [GALLIVAN] and [WEISSTEIN].)

Each year for the past six years at St. Mark's School students attempted to break this record by going for a 13th fold. They never succeeded. Nonetheless, in December of 2011, with the support of OrigaMIT, 17 students went to MIT's famous "infinite corridor" armed with 54,000 feet of paper and managed to construct an object that is physically identical to a strip of paper folded in half 13 times from a single direction, right to left each time. This does not break Gallivan's record as she defines if they taped 4000 foot strips together and layered paper in clever ways during the taping process but it is a significant accomplishment in its own right. As for as they are aware, this is the only example of unidirectional 13-folded construct of some kind made with paper.

References

[ATKINSON1] Atkinson, M.D., On zigzag permutations and comparisons of adjacent elements, *Information Processing Letters*, **21** (1985), 187–189.

[ATKINSON2] Atkinson, M.D., On computing the number of linear extensions of a tree, *Order*, **7** (1990), 23–25.

[BRIGHTWELL and WINKLER] Brightwell G. and Winkler, P., Counting linear extensions, *Order*, **8** (1991), 225–242.

[DAVIS and KNUTH] Davis, C. and Knuth D.J., Number representations and dragon curves, *Journal of Recreational Mathematics,* **3** (1970) 66–81 (Part 1), 133–149 (Part 2).

[GALLIVAN] Gallivan, B. C., *How to Fold Paper in Half Twelve Times: An "Impossible Challenge" Solved and Explained*. Historical Society of Pomona Valley. Pomona, CA, 2002.

[MILLAR, SLOANE and YOUNG] Millar, J., Sloane N.J., and Young, N.E., A new operation on sequences: the Boustrophedon transform, *Journal of Combinatorial Theory, Series A,* **76** (1996), 44–54.

[RIDDLE] Riddle, L., Tiling the plane with the Lévy dragon. URL: http://ecademy.agnesscott .edu/~ lriddle/ifs/levy/tiling.htm

[TOILETPAPERWORLD] Toiletpaperworld Blog: "Beating a paper-folding record with toilet paper." April 14, 2011. URL: http://blog.toiletpaperworld.com

[WEISSTEIN] Weisstein, E. W., Folding. From *MathWorld*—A Wolfram Web Resource. URL: http://mathworld.wolfram.com/Folding.html

9

Folding and Pouring

PUZZLER

One gallon of water is distributed between two containers labeled A and B. Three-quarters of the contents of A are poured into B, and then half the contents of B are poured back into A.

This process of alternately pouring from A to B (three-quarters of the content) and then from B to A (half the content) is repeated.

What happens in the long run?

TIDBIT: Paper Folding

Here's something to try:

> Take a strip of paper and make a crease mark at an arbitrary position.
>
> Make a new crease halfway between the position and the left end of the strip by folding the left end of the paper.
>
> Make a new crease halfway between the new mark and the right end of the strip by folding the right end of the paper.
>
> Repeat, alternating left and right folds, with each fold made to the most recent crease mark.
>
> The system seems to converge to two positions on the strip. What are they?

To answer.... Suppose the strip is 1 unit long and the first crease is at position x. Then a left fold creates a new crease at position

$$\frac{x}{2}$$

and a right fold a crease at position

$$x + \frac{1-x}{2} = \frac{1}{2} + \frac{x}{2}.$$

These transformations have particularly nice interpretations if we write x in base two.

Aside on Base Two. In base ten arithmetic the decimal $0.abcd\ldots$ represents $\frac{a}{10} + \frac{b}{100} + \frac{c}{1000} + \frac{d}{10000} + \cdots$. In base two $0.abcd\ldots$ represents $\frac{a}{2} + \frac{b}{4} + \frac{c}{8} + \frac{d}{16} + \cdots$.

Every number has a base two representation. (Is this obvious?) For example, $3/4 = 0.11$, $1/3 = 0.010101\ldots = 1$. (Why? Read on!) If

$$x = \frac{a}{2} + \frac{b}{4} + \frac{c}{8} + \frac{d}{16} + \cdots = 0.abcd\ldots$$

then

$$2x = a + \frac{b}{2} + \frac{c}{4} + \frac{d}{8} + \cdots = a.bcd\ldots$$

and

$$\frac{x}{2} = \frac{a}{4} + \frac{b}{8} + \frac{c}{16} + \frac{d}{32} + \cdots = 0.0abcd\ldots$$

This shows that *multiplying and dividing by two shifts the decimal point.*

So ... To evaluate $w = 0.111111\ldots$

$$2w = 1.1111\ldots$$
$$= 1 + 0.111\ldots \quad = 1 + w,$$

giving $w = 1$. Also, if $z = 0.010101\ldots$ then $2z + z = 0.11111\ldots = 1$ yielding $z = 1/3$.

Exercise. What is the value of $0.001001001\ldots$ in base two? What is $0.090909\ldots$ in base 10?

Now back to paper folding ...

If the initial crease is at $x = 0.abcd\ldots$ then a left fold produces a new crease at

$$x/2 = 0.0abcd\ldots$$

and a right fold at

$$\frac{1}{2} + \frac{x}{2} = 0.1 + 0.0abcd\ldots = 0.1abcd\ldots$$

These insert a 0 or a 1 in the first slot of the binary representation of x. Thus if we make four right and left folds, we'll have a crease at position $0.0101abcd\ldots$, or with ten folds at position $0.0101010101abcd\ldots$ and so on. After more folds we obtain creases closer and closer to position $0.01010101\ldots = 1/3$ and to position $0.101010101\ldots = 2/3$. (Did you see this on your paper?) The location of the initial fold is irrelevant. The folds always converge to these two locations.

Comment. Paper folding is equivalent to a water-transfer problem: *One gallon of water is distributed between two containers labeled A and B. Half the contents of A are poured into B (changing contents of A by $x \mapsto \frac{x}{2}$) and then half the contents of B are poured back into A (changing contents of A by $x \mapsto x + \frac{1-x}{2} = \frac{1}{2} + \frac{x}{2}$). The process of alternately pouring half from A to B and then half from B to A is repeated. What happens in the long run?*

Answer. The system approaches the same one-third/two-third oscillation.

Instead of transferring half the contents according to the pattern $A \to B$, $B \to A$ what if we transferred half following the cycle $A \to B$, $A \to B$, $B \to A$? (This corresponds to transferring three-quarters of the contents of A to B and bringing half of the contents back. This is the opening puzzler of the newsletter.) Do you see that this converges to the state of having $0.100100100\cdots = 4/7$ gallons in container A at the start of each cycle (followed by 2/7, then 1/7, and back to 4/7)?

Another variation: Suppose we transfer $\frac{9}{10}$ of the liquid back and forth between the two containers according to $A \to B,\ B \to A$. The transformations

$$x \mapsto \frac{x}{10} \qquad x \mapsto \frac{9}{10} + \frac{x}{10}$$

have nice interpretations in base 10 decimals. Can you see that this converges to the oscillatory state of $0.090909\cdots = 1/11$ and $0.909090\cdots = 10/11$ gallons?

Research Corner.   Consider a discrete version of this game.

Suppose we have 14 marbles in one cup labeled A and 18 marbles in a second cup labeled B.

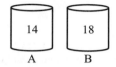

Pour half the contents of cup A into B . . .

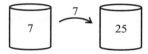

. . . and then half the contents of cup B back into A keeping the extra odd marble in B.

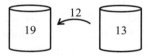

Let's make the rule: *Cup B always keeps or is given any extra odd marble.* Repeatedly pouring half (or just over half in the odd case) of the contents of A into B and then half (or just under half in the odd case) of the contents of B into A enters a 10/22–21/11 oscillation:

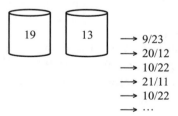

a) Does every initial distribution of 32 marbles lead to a 10/22–21/11 oscillation?

b) Prove that if the number of marbles with which we start is a power of two, then every initial distribution enters the same oscillation. Show that this is not the case for the 9 marble game.

d) Is there a general theory about the types of oscillations to which games will converge? Does there exist a game that enters a cycle with period different from two?

e) How does this change if we handle the odd marbles in a different way?

f) What if we transferred fractions different from one half?

COMMENTARY, SOLUTIONS and THOUGHTS

Folding dyadic fractions made an appearance in newsletter 7. Their reappearance here in connection with water pouring is nice.

Let's examine pouring in more detail. Rather than ask what results from performing a prescribed series of pouring actions, let's ask if we can find a set of pouring actions that yield a prescribed result!

Here's the set-up:

We have one gallon of water distributed between two containers A and B, starting with x gallons in A and $1 - x$ gallons in B. We have a number r between 0 and 1, and we are allowed to pour that proportion of the contents of A into B, or vice versa.

We ask:

Given a value $\alpha \in (0, 1)$, is there a sequence of pouring moves that would converge to α gallons of water in container A?

Pouring half. $r = \frac{1}{2}$

As we saw in the newsletter, if we write $x = 0.abcd \ldots$ as a binary decimal, pouring half the water from container A to B leaves $\frac{x}{2} = 0.0abcd \ldots$ gallons in A (insert 0), and the act of pouring half the contents of B into A leaves $\frac{x}{2} + \frac{1}{2} = 0.1abcd \ldots$ gallons in A (insert 1). If we approximate α by truncating its binary decimal expansion at the nth decimal place and follow the sequence of pouring instructions given by the first n digits after the decimal point, we obtain an amount of water in container A that matches α to n places. It differs from α gallons by no more than $\frac{1}{2^{n+1}}$ gallons, which can be made arbitrarily small by choosing n sufficiently large. (Of course, if α is a dyadic fraction, we can match this amount precisely.)

Pouring More than Half: $r = \frac{2}{3}$

Suppose we now pour two-thirds of the water from one container to the other at each move. If there are x gallons in container A, pouring from A to B leaves $\frac{x}{3}$ gallons in this container, and pouring from B to A produces $x + \frac{2}{3}(1 - x) = \frac{x}{3} + \frac{2}{3}$ gallons in A. If we write x as a decimal in base 3, $x = 0.abcd \ldots$ they correspond to

$$0.abcd \ldots \mapsto 0.0abcd \ldots$$
$$0.abcd \ldots \mapsto 0.2abcd \ldots.$$

Thus we can approximate any quantity α within container A arbitrarily closely provided that α has a ternary expansion involving only the digits 0 and 2. That is, we can approach any value in Cantor's middle-thirds set but we cannot reach values that are not in that set! (For instance, we will never see half a gallon of water in container A with close accuracy, unless we start with that amount in A!)

Challenge. Let r be any value between $\frac{1}{2}$ and 1. Write $r = 1 - \frac{1}{b}$ with $b > 2$. Then a pouring action changes the amount x of water in container A to either $\frac{x}{b}$ gallons or to $\frac{x}{b} + \frac{b-1}{b}$ gallons. Thinking of decimals in base b (even if b is not an integer?) what can we say about which quantities α can be approximated as an amount of water in A?

Pouring Less than Half: $r = \frac{1}{3}$

Suppose we now pour one-third of the water from one container to the other at each move. If there are x gallons in container A, pouring from A to B leaves $\frac{2}{3}x = \frac{x}{3/2}$ gallons in this container, and pouring from B to A produces $x + \frac{1}{3}(1 - x) = \frac{x}{3/2} + \frac{1/2}{3/2}$ gallons in A. If we write x as a decimal in base $1\frac{1}{2}$ using the digits 0 and $1/2$, $x = 0.abcd\ldots$, they correspond to

$$0.abcd\ldots \mapsto 0.0abcd\ldots$$
$$0.abcd\ldots \mapsto 0.\tfrac{1}{2}abcd\ldots.$$

Comment. This is strange! By the "decimal $0.abcd\ldots$ in base one-and-a-half" we mean the number in the interval $[0, 1]$ given by the sum $a\left(\frac{2}{3}\right) + b\left(\frac{2}{3}\right)^2 + c\left(\frac{2}{3}\right)^3 + d\left(\frac{2}{3}\right)^4 + \cdots$ with each coefficient being either 0 or $\frac{1}{2}$. Any number x in the unit interval can be approximated by a finite decimal in base one-and-a-half:

If $\frac{1}{2}\left(\frac{2}{3}\right)$ is smaller than x, choose $a = \frac{1}{2}$, otherwise set $a = 0$.

If $\frac{1}{2}\left(\frac{2}{3}\right)^2$ is smaller than $x - a\left(\frac{2}{3}\right)$, choose $b = \frac{1}{2}$, otherwise set $b = 0$.

If $\frac{1}{2}\left(\frac{2}{3}\right)^3$ is smaller than $x - a\left(\frac{2}{3}\right) - b\left(\frac{2}{3}\right)^2$ choose $c = \frac{1}{2}$, otherwise set $c = 0$, and so on.

If we do this n times, then the sum of the first n terms differs from x by no more than $\frac{1}{2}\left(\frac{2}{3}\right)^{n+1} + \frac{1}{2}\left(\frac{2}{3}\right)^{n+2} + \cdots = \frac{1}{2}\left(\frac{2}{3}\right)^{n+1} \cdot 3 = \left(\frac{2}{3}\right)^n$ which can be made arbitrarily small.

Question. I was careful to avoid the issue of writing x as an infinite decimal. Can this construction be followed to an infinite degree? Does every quantity in $[0, 1]$ have a (well defined?) base one-and-a-half decimal expansion?

For any α we now see that there is a sequence of pouring moves that produces that amount in container A up to any prescribed degree of accuracy. (Write α in base one-and-a-half and follow the sequence of pouring moves they dictate.)

Challenge. Let r be between 0 and $\frac{1}{2}$. Prove that we can approximate any amount $\alpha \in (0, 1)$ gallons of water in container A to any degree of accuracy, with no restriction on the value of α.

These challenges are discussed and solved in [IGA] in a different way.

Discrete Pouring. We can answer challenge b) posed in the research corner relatively easily:

Two containers A and B contain a total of 2^n marbles between them. We pour half the marbles from A into B (taking the extra marble as well into B if we have an odd count) and then half the marbles from B into A (leaving the extra marble in B if we have an odd count). These pouring actions from A to B and then from B to A are repeated indefinitely. What happens in the long run?

If there are x marbles in container A, pouring from A to B leaves $\lfloor \frac{x}{2} \rfloor$ in it and a little thought shows that pouring the contents from B to A yields $\lfloor \frac{x}{2} \rfloor + 2^{n-1}$ marbles in container A. If we write x as an n-digit number in base two, $x = abc\ldots de$, then the operations correspond to deleting the last digit and inserting a 0 and a 1, respectively, in the leftmost position:

$$x = abc\ldots de \rightarrow \left\lfloor \tfrac{x}{2} \right\rfloor = 0abc\ldots d$$
$$x = abc\ldots de \rightarrow \left\lfloor \tfrac{x}{2} \right\rfloor + 2^{n-1} = 1abc\ldots d.$$

Thus, eventually, the contents of container A oscillates between two values: $1010101\ldots = \left\lfloor \tfrac{2}{3} \cdot 2^n \right\rfloor$ and $0101010\ldots = \left\lfloor \tfrac{1}{3} \cdot 2^n \right\rfloor$.

Reference

[IGA] Iga, K., The truck driver's straw problem and Cantor sets, *College Mathematics Journal*, **39** (2008), 280–290.

10 Fractions

PUZZLER

a) The following are true:

$$\frac{26}{65} = \frac{2}{5} \qquad \frac{266}{665} = \frac{2}{5} \qquad \frac{2666}{6665} = \frac{2}{5}.$$

Does 2 followed by *n* sixes divided by *n* sixes followed by 5 always equals 2/5?

b) We also have

$$\frac{49}{98} = \frac{4}{8} \qquad \frac{499}{998} = \frac{4}{8} \qquad \frac{4999}{9998} = \frac{4}{8}.$$

Does 4 followed by *n* nines divided by *n* nines followed by 8 always equals 4/8?

c) Find another example.

Egyptian Fractions

(This material is adapted from Chapter 11 of *THINKING MATHEMATICS! Volume 1: Arithmetic = Gateway to All.*)

Suppose we wish to share 7 pies among 12 boys.

We could divide each pie into 12 parts and give each boy 7 pieces. But let's be practical as were the ancient Egyptians (ca. 2000 B.C.E.). Here are the seven pies:

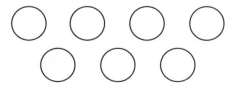

Is it possible to give each boy a whole pie? No. How about the next best thing—each boy half a pie? Yes! There are 12 half pies to distribute. There is one pie left to be shared among the 12 boys. Divide it into twelfths and give each boy an extra piece.

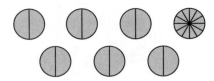

Thus each boy receives $\frac{1}{2} + \frac{1}{12} = \frac{7}{12}$ of a pie.

Unit fractions, that is, fractions with 1 in the numerator, are easy to work with in practice and we have a good intuitive feel for them. Egyptian tradesmen and scholars expressed all fractions as sums of *distinct* unit fractions. For example, 3/10 was written $\frac{1}{4} + \frac{1}{20}$ and 5/7 as $\frac{1}{2} + \frac{1}{5} + \frac{1}{70}$. (Thus, to share 3 pies among 10 students, give each student one-quarter and one-twentieth of a pie.) The Egyptians placed a dot above a number to represent the corresponding unit fraction. Thus $5/7 = \dot{2} + \dot{5} + 7\dot{0}$. They expressed all fractions as sums in this way, except for the fraction 2/3 which had its own symbol.

Question. Why did the Egyptians insist on writing fractions as sums of distinct unit fractions and not allow repetition of terms? (For instance, $3/10 = \dot{4} + 2\dot{0}$ is preferred over $3/10 = 1\dot{0} + 1\dot{0} + 1\dot{0}$.) Could it be a practical concern? For instance, to share 3 pies among 10 students we could divide each pie into tenths and give each student three parts. Is receiving two parts (namely, a quarter and a twentieth), with one piece fairly large, preferable? What do you think? Care to conduct some internet research on the topic?

The Egyptians were adept at computing sums of unit fractions. It is typically not easy to do.

Example. Write $\frac{4}{13}$ as a sum of distinct unit fractions.

Answer. The Egyptians preferred always to take out the largest unit fraction possible. Writing $\frac{4}{13} = \frac{1}{3\frac{1}{4}}$, it is clear that $\frac{1}{3}$ is larger than $\frac{4}{13}$, but $\frac{1}{4}$ is smaller. With some scratch-work we see $\frac{4}{13} = \frac{1}{4} + \frac{3}{52}$. Now $\frac{3}{52} = \frac{1}{17\frac{1}{3}}$, which shows that $\frac{1}{18}$ is the next largest unit fraction we can take out. We have $\frac{3}{52} - \frac{1}{18} = \frac{1}{468}$ and so $\frac{4}{13} = \frac{1}{4} + \frac{1}{18} + \frac{1}{468}$. We're done!   □

Exercise. Write 3/7 as a sum of distinct unit fractions by removing at each stage the largest unit fraction. Do the same for 5/11.

In 1202 Italian scholar Fibonacci questioned whether every fraction can be written as a sum of distinct unit fractions. Does the process of taking out the largest unit fraction always yield the desired outcome? He showed that it did.

Challenge: Fibonacci's Proof

Consider a fraction $\frac{a}{b}$ with $a < b$.

a) Let N be the smallest positive integer with $\frac{a}{b} > \frac{1}{N}$, so $\frac{a}{b} < \frac{1}{N-1}$. Write $\frac{a}{b} - \frac{1}{N}$ over a common denominator and show that it has a numerator that is positive and smaller than a.

b) Explain why if we repeat this process we eventually obtain a fraction with numerator 1.

c) Explain why $\frac{a}{b}$ is then a sum of (distinct) unit fractions.

Exercise. Write the fraction $\frac{12}{17}$ as a sum of distinct unit fractions two different ways.

Challenge. Use Fibonacci's method to write the fraction $\frac{1}{1}$ as an *infinite* sum of distinct unit fractions. What can be said about the denominators that appear?

Research Corner. Galilleo Galilei (1564–1642) observed that the odd numbers have a curious property:

$$\frac{1}{3} = \frac{1+3}{5+7} = \frac{1+3+5}{7+9+11} = \frac{1+3+5+7}{9+11+13+15} = \cdots .$$

This April, students of the Math Institute "Math Chat" class came up with a visual proof:

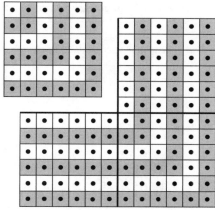

$$\frac{1+3+5+7+9+11}{13+15+17+19+21+23} = \frac{1}{3}$$

$$\frac{\text{The sum of the first } k \text{ odds}}{\text{The sum of next } k \text{ odds}} = \frac{1}{3}$$

This generalizes:

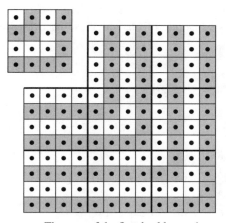

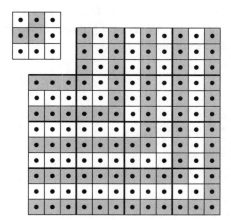

$$\frac{\text{The sum of the first } k \text{ odds}}{\text{The sum of the next } 2k \text{ odds}} = \frac{1}{8}$$

$$\frac{\text{The sum of the first } k \text{ odds}}{\text{The sum of the next } 3k \text{ odds}} = \frac{1}{15}$$

$$\frac{\text{The sum of the first } k \text{ odds}}{\text{The sum of next } mk \text{ odds}} = \frac{1}{(m+1)^2 - 1}$$

Can you discover something analogous about the even numbers? About other arithmetic sequences?

COMMENTARY, SOLUTIONS and THOUGHTS

In one of my courses for teachers we discuss the mathematics of fractions. I present the following little known "fact." (See [TANTON, Chapter 11].)

It is possible to cross out any repeated 3s, 6s, and 9s that appear in a fraction and not change its value.

For example,

$$\frac{2\cancel{6}}{\cancel{6}5} = \frac{2}{5} \qquad\qquad \frac{1\cancel{9}}{\cancel{9}5} = \frac{1}{5}$$

$$\frac{1\cancel{6}}{\cancel{6}4} = \frac{1}{4} \qquad\qquad \frac{2\cancel{6}\cancel{6}\cancel{6}}{\cancel{6}\cancel{6}65} = \frac{2}{5}$$

$$\frac{4\cancel{9}}{\cancel{9}8} = \frac{4}{8} \qquad\qquad \frac{16\cancel{3}}{\cancel{3}26} = \frac{1}{2}$$

The opening puzzler of the newsletter offers more examples.

The "fact" is false and the examples are coincidental. (Clearly $\frac{23}{37}$ is not $\frac{2}{7}$, for instance.) It is an interesting challenge to find other examples where such anomalous cancellation happens to hold. John McCarthy, an avid follower of the St. Mark's Institute of Mathematics, discovered this impressive example:

$$\frac{12345679}{98765432} = \frac{1}{8}$$

along with others

$$\frac{99}{198} = \frac{9}{18} \qquad \frac{495}{693} = \frac{45}{63}$$

$$\frac{34}{136} = \frac{4}{16} \qquad \frac{436}{763} = \frac{4}{7}$$

$$\frac{22}{121} = \frac{2}{11} \qquad \frac{138}{1840} = \frac{3}{40}$$

$$\frac{67}{469} = \frac{7}{49} \qquad \frac{34}{238} = \frac{4}{28}$$

St. Mark's student Swetha Dravida '09, discovered that the example $\frac{26}{65} = \frac{2}{5}$ generalizes to any equal number of 6s in the numerator and denominator:

$$\frac{2}{5} = \frac{26}{65} = \frac{266}{665} = \frac{2666}{6665} = \cdots$$

To see why this holds think of numbers in terms of counts of units, tens, hundreds, and so on. Then

$$266 \cdots 66 \times 5 = 10|30|30| \cdots |30|30$$
$$= 10|30|30| \cdots |33|0$$
$$= 13|3|3| \cdots |3|0$$

and

$$66 \cdots 665 \times 2 = 12|12| \cdots |12|12|10$$
$$= 12|12| \cdots |12|13|0$$
$$= 12|12| \cdots |13|3|0$$
$$= 13|3| \cdots |3|3|0,$$

which are equal.

In the same way, $499 \cdots 99 \times 8 = 99 \cdots 998 \times 4$ and $199 \cdots 99 \times 5 = 99 \cdots 995$ giving

$$\frac{1}{5} = \frac{19}{95} = \frac{199}{995} = \cdots .$$

For more examples of anomalous cancellation see [BOAS] and [WEISSTEIN].

Egyptians Fractions

The newsletter offers only a brief introduction to this rich and mathematically vast topic. Ron Knott [KNOTT] has a particularly impressive, and interactive, webpage outlining the history, use, and mathematics of unit fractions.

It is true that Fibonacci's "greedy" algorithm described in the newsletter of always choosing the largest possible workable fraction first will convert any fraction $\frac{a}{b}$ (with $a < b$) into a sum of distinct unit fractions.

Let N be the smallest positive integer with $\frac{a}{b} > \frac{1}{N}$, so $\frac{a}{b} < \frac{1}{N-1}$. Then $\frac{a}{b} - \frac{1}{N} = \frac{Na-b}{bN}$ is a fraction with positive numerator (because $\frac{a}{b} > \frac{1}{N}$), smaller than the original numerator a (because $\frac{a}{b} < \frac{1}{N-1}$), and with a larger denominator. Repeated application eventually produces a fraction with numerator 1. When this happens we have an expression of the form

$$\frac{a}{b} - \frac{1}{N_1} - \frac{1}{N_2} - \cdots - \frac{1}{N_k} = \frac{1}{M},$$

allowing us to write $\frac{a}{b}$ as a sum of distinct unit fractions.

Representations are not unique. For example, $\frac{12}{17} = \frac{1}{2} + \frac{1}{5} + \frac{1}{170} = \frac{1}{2} + \frac{1}{6} + \frac{1}{30} + \frac{1}{170}$. Given one representation of a fraction as a sum of k unit fractions, using the identity

$$\frac{1}{N} = \frac{1}{N+1} + \frac{1}{N(N+1)}$$

on the final term converts the representation into a sum of $k + 1$ unit fractions.

For each fraction there is a shortest length for its representation as an Egyptian fraction. Even the shortest representations need not be unique. For example

$$\frac{3}{7} = \frac{1}{3} + \frac{1}{11} + \frac{1}{231} = \frac{1}{4} + \frac{1}{6} + \frac{1}{84}$$

Challenge. Show that $\frac{3}{7}$ cannot be the sum of two distinct unit fractions.

Basic questions about the shortest representations of fractions remain unsolved. For example, it is conjectured, but not proved, that every fraction of the form $\frac{4}{n}$ can be written as a sum of three unit fractions. (This is known as the *Erdős-Straus Conjecture*. See [GUY].)

Consider again the identity

$$\frac{1}{N} = \frac{1}{N+1} + \frac{1}{N(N+1)}.$$

Applying it to $\frac{1}{1}$ produces the sequence

$$\frac{1}{1} = \frac{1}{2} + \frac{1}{2}$$
$$\frac{1}{1} = \frac{1}{2} + \frac{1}{3} + \frac{1}{6}$$
$$\frac{1}{1} = \frac{1}{2} + \frac{1}{3} + \frac{1}{7} + \frac{1}{42}$$
$$\frac{1}{1} = \frac{1}{2} + \frac{1}{3} + \frac{1}{7} + \frac{1}{43} + \frac{1}{1806}$$
$$\cdots$$

This gives a representation of 1 as an infinite sum of distinct unit fractions:

$$1 = \frac{1}{2} + \frac{1}{3} + \frac{1}{7} + \frac{1}{43} + \frac{1}{1807} + \frac{1}{3263443} + \cdots$$

Challenge. Each denominator is the product of the previous denominators plus one. Does this continue? Are denominators always coprime?

Challenge. We have

$$-\frac{1}{2} + \frac{1}{3} = -\frac{1}{2} \cdot \frac{1}{3}$$
$$-\frac{1}{2} + \frac{1}{3} + \frac{1}{7} = -\frac{1}{2} \cdot \frac{1}{3} \cdot \frac{1}{7}$$
$$-\frac{1}{2} + \frac{1}{3} + \frac{1}{7} + \frac{1}{43} = -\frac{1}{2} \cdot \frac{1}{3} \cdot \frac{1}{7} \cdot \frac{1}{43}$$

Does this pattern persist?

Comment. There is more than one way to write 1 as an infinite sum of distinct unit fractions. For example:

$$1 = \frac{1}{2} + \frac{1}{4} + \frac{1}{8} + \frac{1}{16} + \cdots.$$

Final Comment. The work of St. Mark's Institute students mentioned in the newsletter appears in more detail in Appendix V.

References

[BOAS] Boas, R. P., Anomalous cancellation in *Mathematical Plums* (ed. R. Honsberger), Mathematical Association of America, Washington D.C., 1979.

[GUY] Guy, R., *Unsolved Problems in Number Theory, 2nd ed.*, Springer-Verlag, New York, NY, 1994.

[KNOTT] Knott, R., Egyptian fractions. URL: http://www.maths.surrey.ac.uk/hosted-sites/R.Knott/Fractions/egyptian.html

[TANTON] Tanton, J., *Thinking Mathematics! Volume 1: Arithemetic = Gateway to All*, www.lulu.com, 2009.

[WEISSTEIN] Weisstein, E.W., Anomalous cancellation from *MathWorld*-A Wolfram Web Resource. URL: http://mathworld.wolfram.com/AnomalousCancellation.html

Integer Triangles

PUZZLER: More than just an Area!

What property do each of the following figures share?

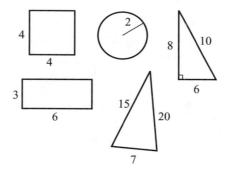

Find another right triangle (with integer sides) with the property. Is there another integer rectangle with the property?

Heron's Formula. In 100 C.E. Heron of Alexandria (also known as Hero of Alexandria) published a remarkable formula for the area of a triangle in terms of its three side-lengths

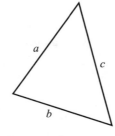

$$\text{area} = \frac{1}{4}\sqrt{(a+b+c)(a+b-c)(a-b+c)(-a+b+c)}$$

77

Thus the area A of the 15-20-7 triangle above is

$$A = \frac{1}{4}\sqrt{(42)(2)(28)(12)} = \frac{\sqrt{28224}}{4} = 42.$$

Proving Heron's formula is not difficult conceptually. (The algebra required, on the other hand, is a different matter!) Here are two possible approaches:

Proof 1. Draw an altitude and label the lengths x, $b - x$, and h as shown:

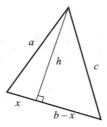

We have

$$(b - x)^2 + h^2 = c^2,$$
$$x^2 + h^2 = a^2.$$

Subtract to obtain a formula for x and substitute to obtain a formula for h. Use $A = \frac{1}{2}bh$ (and three pages of algebra) to obtain Heron's formula.

Proof 2. We have $A = \frac{1}{2}ab \sin \theta$.

By the law of cosines $c^2 = a^2 + b^2 - 2ab \cos \theta$. Solve for $\sin \theta$ and for $\cos \theta$ and substitute into $\cos^2 \theta + \sin^2 \theta = 1$.

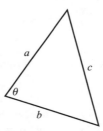

Out (eventually!) pops Heron's formula.

Toothpick Triangles. Here's a great student activity:

With 11 toothpicks it is possible to make four different triangles of perimeter 11 using a whole number of toothpicks per side

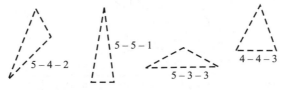

(Why isn't 6-3-2 a valid triangle?)

*The count goes down if one adds another toothpick to the mix: With 12 toothpicks we can make only **three** different triangles: 5-5-2, 5-4-3 and 4-4-4.*

What's going on?

Construct a table showing the number of integer triangles we can make with $1, 2, \ldots, 20$ toothpicks. What do you notice about the even and odd entries? Do you see any patterns?

In 2005, high-school students (and some younger) of the St. Mark's Research Group played with this problem and discovered—and proved—the following remarkable formula:

The number of triangles that can be made with N toothpicks is $\left\langle \frac{N^2}{48} \right\rangle$ if N is even and $\left\langle \frac{(N+3)^2}{48} \right\rangle$ if N is odd.

(The angled brackets mean "nearest integer to.")

Thus we can make $\left\langle \frac{100^2}{48} \right\rangle = 208$ different triangles with 100 toothpicks!

Research Corner: Going for Integers all Round. The 15-20-7 triangle on the previous page shows that it is possible for a triangle to have integer side lengths and integer area.

There are many examples. Any right triangle with one leg an even integer has integral area:

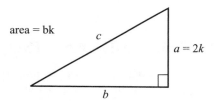

But the 15-20-7 triangle has the added property that, not only are its side lengths, perimeter, and area integers, the perimeter and the area have the same value, namely, 42.

Challenge 1. Find another (non-right) integer triangle with perimeter equal to area.

Another idea ... It is possible for two different integer triangles to have the same perimeter and the same area? The following two triangles are an example.

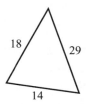 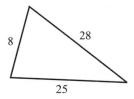

Alas, their areas are not integers.

Challenge 2. Does there exist a pair of integer triangles with the same perimeter and the same integer area?

Challenge 3. Let $T(N)$ be the number of integer triangles with perimeter N and integral area. Is there a formula for $T(N)$ akin to the formula for toothpick triangles?

Just for fun . . . Here's an integer tetrahedron with each face of integer area and volume also an integer!

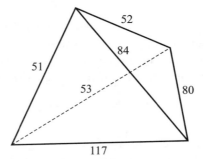

COMMENTARY, SOLUTIONS and THOUGHTS

Each shape in the opening puzzler of the newsletter has the property that the numerical value of its perimeter equals the numerical value of its area. (Is one allowed to say "perimeter equals area"?)

There are only two integer rectangles with this property.

Suppose a rectangle with sides of lengths a and b has area matching perimeter. Then $ab = 2a + 2b$ and so

$$b = \frac{2a}{a-2} = 2 + \frac{4}{a-2}.$$

For the right-hand side to be an integer $a - 2$ must be 1, 2, or 4. Thus only the 3×6 rectangle and the 4×4 square have the property.

Comment. As observed by high school teacher Michael Ericson, the condition $ab = 2a + 2b$ can be written $\frac{1}{a} + \frac{1}{b} = \frac{1}{2}$. (Read on!)

There are only two integer right triangles with the property:

Suppose a right triangle with legs of integer lengths a and b has area matching perimeter. Then $\frac{1}{2}ab = a + b + \sqrt{a^2 + b^2}$. Squaring

$$a^2 + b^2 = \left(\frac{1}{2}ab - a - b\right)^2 = \frac{1}{4}a^2b^2 + a^2 + b^2 - a^2b - ab^2 + 2ab$$

so

$$a^2b + ab^2 = \frac{1}{4}a^2b^2 + 2ab.$$

Dividing by ab and solving for b gives:

$$b = \frac{4a - 8}{a - 4} = \frac{4a - 16 + 8}{a - 4} = 4 + \frac{8}{a - 4}.$$

For the right-hand side to be an integer, a must be 5, 6, 8, or 12. This shows that the 6-8-10 and 5-12-13 triangles are the only integer right triangles with the desired property.

If we remove the requirement that the side lengths of the triangle are integral (and consider arbitrary triangles), we can say:

The perimeter and area of a triangle have the same numerical value if and only if the inradius of the triangle is 2.

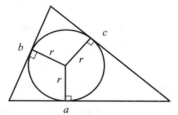

Following the notation in the diagram we see that the area A and perimeter P of a triangle satisfy

$$A = \frac{1}{2}ar + \frac{1}{2}br + \frac{1}{2}cr = \frac{1}{2}Pr.$$

(The same relation holds for a polygon with sides tangent to a common circle.)

Comment. The condition that the inradius r be two can be written $\frac{1}{r} = \frac{1}{2}$.

Curious Challenge

a) Show that a parallelogram has area equal to perimeter if, and only if, $\frac{1}{h_1} + \frac{1}{h_2} = \frac{1}{2}$ where h_1 and h_2 are its two heights.

b) Show that the area of a triangle equals its perimeter if and only if $\frac{1}{h_1} + \frac{1}{h_2} + \frac{1}{h_3} = \frac{1}{2}$, where h_1, h_2, and h_3 are the three altitudes of the triangle.

c) Show that an $a \times b \times c$ rectangular prism has volume equal to its surface area if and only if $\frac{1}{a} + \frac{1}{b} + \frac{1}{c} = \frac{1}{2}$.

d) Show that a right circular cylinder has volume equal to surface area if and only if $\frac{1}{r} + \frac{1}{h} = \frac{1}{2}$, where r is the radius of its base and h is its height.

e) Show that a regular circular cone has volume equal to surface area if and only if $\frac{1}{r} + \frac{1}{b} = \frac{1}{2}$ where r is the radius of its base and b is the perpendicular distance from the center of its base to its lateral surface.

f) Is something going on here? Can we associate with any geometric figure some fundamental set of (perpendicular?) measurements that dictate the properties of its area and perimeter or, in three dimensions, its volume and its surface area? The volume of a sphere equals surface area if and only if $\frac{1}{r} = \frac{1}{3}$. Would it be more appropriate to write $\frac{1}{d} + \frac{1}{d} + \frac{1}{d} = \frac{1}{2}$ where d is the diameter?

See the comment at the end of this essay for a few thoughts on this.

Pairs of Integer Triangles of Equal Area and Equal Perimeter

In the newsletter I presented a pair of integer triangles with the same perimeter and the same area:

The 14-18-29 *and* 8-25-28 *triangles each have perimeter 61 and area* $\frac{5\sqrt{3111}}{4}$.

As is often the case with such problems, if one example exists, infinitely many do! Tyler Jarvis of Brigham Young University pointed me in the right direction to hunt for other solutions.

The challenge is to find integer solutions to

$$a + b + c = d + e + f,$$
$$s(s-a)(s-b)(s-c) = s(s-d)(s-e)(s-f),$$

where $s = \frac{a+b+c}{2}$. (Here a, b, c and d, e, f represent the side-lengths of the two triangles.) Eliminating f and simplifying means we seek integer solutions to

$$(-a+b+c)(a-b+c)(a+b-c)$$
$$= (a+b+c-2d)(a+b+c-2e)(2d+2e-a-b-c) \qquad (*)$$

Rational solutions will suffice because a rational solution can be converted to an integer solution by multiplication.

Let S denote the set of all rational points (a, b, c, d, e) that satisfy $(*)$.

We have within S a host of trivial solutions, namely those that correspond to the two triangles being identical: $d = a$ and $e = b$ (and consequently $f = c$). Let A be any one of them

$$A = (a, b, c, a, b).$$

We also have a particular solution

$$P = (14,\ 18,\ 29,\ 8,\ 25).$$

Our strategy is to look at the line that connects the point A and the point P and see if it again intersects the set S. For a real number λ, let

$$X_\lambda = \lambda A + (1-\lambda) P$$
$$= (\lambda a + 14(1-\lambda),\ \lambda b + 18(1-\lambda),\ \lambda c + 29(1-\lambda),\ \lambda a + 8(1-\lambda),\ \lambda b + 25(1-\lambda)).$$

For X_λ to lie in S its components must satisfy the equation $(*)$. This yields a cubic in λ. We know that $\lambda = 0$ and $\lambda = 1$ are solutions so we can factor λ and $\lambda - 1$ from it to yield a linear equation in λ with rational coefficients. This gives a third rational solution to $(*)$. We need to check that it corresponds to a meaningful geometric solution (namely, the sides of the triangle are all positive and the triangular inequalities hold so that the triangle actually exists!) but the approach gets us beyond the most difficult part of the search: finding examples of numbers that satisfy $(*)$.

With a computer program we can generate many examples of integer triangle pairs with equal areas and equal perimeters. Here are a few:

$10 - 34 - 39$ and $19 - 24 - 40$
$18 - 27 - 30$ and $20 - 24 - 31$
$35 - 57 - 62$ and $42 - 47 - 65$
$40 - 61 - 66$ and $46 - 52 - 69$
$45 - 94 - 94$ and $49 - 84 - 100$.

Counting Integer Triangles

Students of the St. Mark's Institute Research class played with this challenge in 2005 ([TANTON]), as had the 2002 students of the Boston Math Circle ([FOCUS]).

Counting the number of distinct triangles we can form with $n = 1, 2, 3, \ldots$ toothpicks yields the sequence

$$0, 0, 1, 0, 1, 1, 2, 1, 3, 2, 4, 3, 5, 4, 7, 5, 8, 7, 10, 8, \ldots.$$

Denote the n term by $T(n)$. The list is composed of two intertwined copies of the same sequence of numbers $0, 1, 1, 2, 3, 4, 5, 7, 8, 10, \ldots$, with one copy shifted three places. That is, it seems that

$$T(2n) = T(2n - 3) \text{ for } n > 1.$$

If this is true, we need then only consider triangles with even integer perimeters in our pursuit of a formula for these counts.

Let's take a moment to collate some facts about integer triangles.

1. Three positive integers a, b, and c (written in non-decreasing order) are the side lengths of a triangle if and only if $a + b > c$.

This is the triangle inequality.

2. No integer triangle (or any triangle for that matter) possesses a side of length greater than or equal to half its perimeter.

The remaining two sides would sum to a value less than half the perimeter and this violates Fact 1.

3. No integer triangle with even perimeter has a side of length 1.

This follows from Fact 2.

4. If a, b, and c (written in non-decreasing order) are the sides of an integer triangle of even perimeter, then $a + b > c + 1$.

We know $a + b > c$. If $a + b = c + 1$, then its perimeter $a + b + c$ would be odd.

5. If a, b, and c are the sides of an integer triangle with even perimeter $2n$, then $a - 1$, $b - 1$, and $c - 1$ are the sides of a valid triangle of (odd) perimeter $2n - 3$. Conversely, adding one to the value of each side of a triangle of perimeter $2n - 3$ produces a valid triangle of perimeter $2n$.

One checks that the necessary inequalities hold.

Fact 5 shows that we have a match between the triangles of perimeter $2n$ and those of perimeter $2n - 3$ and so, indeed, $T(2n) = T(2n - 3)$ for all $n > 1$.

In pursuit of a formula for $T(n)$, the 2005 students of the St. Mark's Institute class made the following key discovery:

Lemma. *For N even with $N > 12$, $T(N) - T(N - 12) = \frac{N}{2} - 3$.*

Proof. Write $N = 2n$.

Let $a \leq b \leq c$ represent the sides of an integer triangle of perimeter $N - 12$. By Fact 2, c is at most $n - 7$. Then $a + 4$, $b + 4$, and $c + 4$ represent the side lengths of a valid triangle of perimeter

N with longest side at most $n - 3$. The correspondence

$$(a, b, c) \leftrightarrow (a + 4,\ b + 4,\ c + 4)$$

provides a match between the two types of triangles, but it omits the triangles of perimeter N with longest side $n - 2$ or $n - 1$. How many of these triangles is this correspondence overlooking?

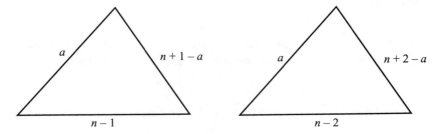

The triangle inequality shows that for a triangle of perimeter $2n$ and longest side $n - 1$, the shortest length a of the triangle satisfies $2 \leq a \leq \frac{n+1}{2}$, and for a triangle of perimeter $2n$ and longest side $n - 2$ the shortest side a satisfies $4 \leq a \leq \frac{n+2}{2}$. There are thus a total of $\lfloor \frac{n+1}{2} \rfloor - 1 + \lfloor \frac{n+2}{2} \rfloor - 3 = n - 3$ of these triangles. Consequently $T(N)$ is larger than $T(N - 12)$ by $n - 3$. This establishes the lemma. $\qquad \square$

Define $T(0)$ to be zero. If we write $N = 12k + r$ with $r = 0, 2, 4, 6, 8$ or 10, then

$$\begin{aligned}
T(N) &= T(N) - T(N - 12) \\
&\quad + T(N - 12) - T(N - 2 \cdot 12) \\
&\quad + \cdots \\
&\quad + T(12 + r) - T(r) \\
&\quad + T(r)
\end{aligned}$$

$$= \frac{12k + r}{2} - 3 + \frac{12(k - 1) + r}{2} - 3 + \cdots + \frac{12 \cdot 1 + r}{2} - 3 + T(r)$$

$$= 6(k + (k - 1) + \cdots + 1) + k \cdot \frac{r}{2} - 3k + T(r)$$

$$= 3k^2 + \frac{1}{2}kr + T(r)$$

$$= \frac{(12k + r)^2}{48} - \frac{r^2}{48} + T(r).$$

We check that $T(r) - \frac{r^2}{48}$ is strictly between $-\frac{1}{2}$ and $\frac{1}{2}$ for each of the six values of r. That is, for each even value N

$$T(N) = \frac{N^2}{48} \pm \varepsilon$$

for some ε less than $1/2$. Thus $T(N) = \left\langle \frac{N^2}{48} \right\rangle$, as claimed in the newsletter. For N odd we have $T(N) = T(N + 3) = \left\langle \frac{(N+3)^2}{48} \right\rangle$.

Challenge. Let $S(n)$ be the count of *scalene* integer triangles of perimeter n. What do you notice about the sequence of numbers produced?

Final Comment on the Area and Perimeter of Polygons.

Suppose a polygon of area A and perimeter P has sides of lengths $a_1, a_2, \ldots, a_k$. Consider the quantity $\frac{P}{2A} = \frac{a_1 + a_2 + \cdots + a_k}{2A}$ (which equals $\frac{1}{2}$ if $P = A$).

Area is the product of two quantities of dimension "length." If there is a natural means to use a sum $a_{i_1} + \cdots + a_{i_n}$ as one of those lengths in the computation of area, then $h = \frac{2A}{a_{i_1} + \cdots + a_{i_n}}$ is a meaningful length in the geometry of the figure and its reciprocal $\frac{1}{h}$ appears in the expression $\frac{P}{2A}$.

For example, for a triangle with sides a, b, and c, its area can be computed as $A = \frac{1}{2}(a + b + c)r$ where r is the inradius of the triangle and

$$\frac{P}{2A} = \frac{1}{r}.$$

Its area can also be computed as $A = \frac{1}{2}ah_a = \frac{1}{2}bh_b = \frac{1}{2}ch_c$ (where h_a, h_b and h_c are the altitudes of the triangle) and

$$\frac{P}{2A} = \frac{a}{2A} + \frac{b}{2A} + \frac{c}{2A} = \frac{1}{h_a} + \frac{1}{h_b} + \frac{1}{h_c}.$$

Does thinking along these lines add to or diminish the idea that something interesting is going on in the Curious Challenge presented earlier in this chapter?

References

[FOCUS] Students of *The Boston Math Circle*, Young students approach integer triangles, *FOCUS*, **22** no. 5 (2002), 4–6.

[TANTON] Tanton, J., Pit your wits against young minds! *Mathematical Intelligencer*, **29** no. 3 (2007), 55–59.

12
Lattice Polygons

This chapter might drive you dotty! As you will see, we're asking a number of tricky questions about polygons on a square lattice of dots. There are many interesting observations to be made about them. Nothing requires high-powered mathematics, just, perhaps, high powered ingenuity! Have a piece of graph paper at your side as you ponder these tricky proposals.

PUZZLER 1: Tricky Triangle

Is it possible to draw an equilateral triangle on a square array of dots so that each corner of the triangle lies on a dot?

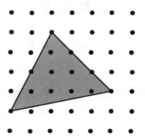

PUZZLER 2: Sneaky Squares

We can draw squares of areas 1, 2, 4, 5, and 8 on a square lattice:

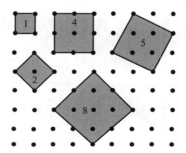

(For example, the square of area five consists of a unit square surrounded by four right triangles, each half a rectangle of area two.) It is also possible to draw squares of areas 9, 10, 13, 16, 17, and 20. Can you see how?

There is one other number N less than 20 for which there is a square of area N. Which number did I skip?

Challenge 1. What property must N have for it to be possible to draw a square of area N on a lattice?

Challenge 2. Prove that the list of numbers 1, 2, 4, 5, 8, 9, 10, . . . is closed under multiplication. That is, if there are lattice squares of areas N and M, then there is a lattice square of area $N \times M$.

PUZZLER 3: Other Equilaterals

A polygon is said to be *equilateral* if its sides all have the same length. For example, here is an equilateral hexagon on a square lattice. (Each side has length five units.)

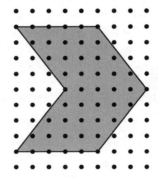

a)   Draw an equilateral lattice octagon.

b)   Explain how we could construct an equilateral lattice polygon with 628 sides.

c)   Draw an example of an equilateral lattice polygon with an odd number of sides.

TIDBIT: Lattice Areas

It is clear that a rectangle drawn on a square lattice with horizontal and vertical sides has area a whole number of units. It is also clear that any right triangle with a horizontal and a vertical side is half a rectangle and so has area that is either half an integer or a whole integer.

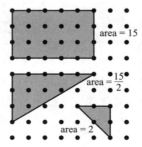

It is surprising that the same result holds true for *any* lattice triangle:

The area of a lattice triangle is either an integer or a half-integer number of units.

The reason follows from a single picture. Surround the triangle with a rectangle yielding three right triangles and possibly another rectangle. (St. Mark's high-school student Nick Roumas noticed this.)

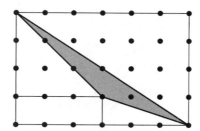

The area of the shaded triangle equals the area of the large rectangle (an integer) minus, possibly, the area of a small rectangle, minus three half or whole integers for the triangles. The arithmetic yields another whole or half-integer.

As any lattice polygon can be subdivided into lattice triangles, this shows

The area of every lattice polygon is either an integer or a half-integer number of units.

This is a special case of a famous result discovered in 1899 by Austrian mathematician Georg Pick (1859–1942):

The area A of a lattice polygon is given by

$$A = I + \frac{B}{2} - 1,$$

where I is the number of lattice points inside the polygon and B is the number of lattice points on the boundary of the polygon.

For example, here's a moderately complex lattice polygon:

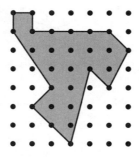

It has $I = 12$ interior points and $B = 15$ boundary points. Its area, as you can check directly, is $A = 12 + 15/2 - 1 = 18\,^1/_2$ square units.

It is easy to see that Pick's formula is valid for rectangles and for right triangles. Using the diagram Nick Roumas studied it is also possible to show that Pick's formula is valid for any lattice triangle. Since every lattice polygon can be decomposed into lattice triangles, it is a hop, skip, and a jump to show that Pick's formula holds for all lattice polygons. (Challenge: This was swift! Can you fill in the details?)

Research Corner. Explore lattice polygons on equilateral triangle lattices.

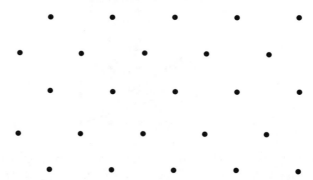

Can we draw a square on such a lattice? Can anything be said about the areas of lattice polygons? Equilateral polygons?

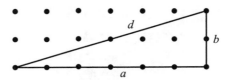

COMMENTARY, SOLUTIONS and THOUGHTS

Lattice Squares

Pythagoras's theorem shows that the length d of a line segment connecting points of a square lattice satisfies $d^2 = a^2 + b^2$ for integers a and b.

Thus the area N of a lattice square with side-length d is a sum of two squares:

$$N = d^2 = a^2 + b^2.$$

Conversely, any N that can be written as a sum of two squares can be the area of a lattice square.

The sequence of numbers that can be so written begins

$$1, 2, 4, 5, 8, 9, 10, 13, 16, 17, 18, 20, \ldots$$

Challenge. The number 25 can be written as a sum of two squares in two different ways: $25 = 5^2 + 0^2$ and $25 = 3^2 + 4^2$, and so it is possible to draw a square of area 25 on a lattice grid in two ways. What is the next number that has two different representations as a sum of two squares? Is there a number that can be represented in three different ways?

Comment. The numbers that are sums of two squares have been completely classified:

N is a sum of two squares if and only if each prime that is one less a multiple of four (3, 7, 11, 19, 23, . . .) that appears in the prime factorization of N does so an even number of times.

For example, $N = 2 \cdot 3^2 \cdot 5^3 \cdot 7^{46} \cdot 13$ is a sum of two squares (it equals $130 \cdot 3^2 \cdot 5^2 \cdot 7^{46} = (11^2 + 3^2) \cdot 3^2 \cdot 5^2 \cdot 7^{46} = (3 \cdot 5 \cdot 7^{23} \cdot 11)^2 + (3^2 \cdot 5 \cdot 7^{23})^2)$ but $M = 2 \cdot 3^2 \cdot 5^3 \cdot 7^{47} \cdot 13$ is not.

We prove this classification result in Appendix I.

The set of integers expressible as a sum of two squares is closed under multiplication:

$$\text{If } N = a^2 + b^2 \text{ and } M = c^2 + d^2, \text{ then}$$

$$NM = (a^2 + b^2)(c^2 + d^2) = (ac + bd)^2 + (ad - bc)^2.$$

This can be established geometrically:

Draw a square of area N on a square lattice. Its vertices lie on a tilted square lattice of dots within the original lattice. Draw a square of area M lattice squares on that tilted lattice. The result is a square on the original lattice of area NM.

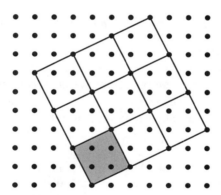

Using a lattice square of area 5 and a lattice square of area 9 to make a square of area 45.

Equilateral Figures

The newsletter establishes that every lattice triangle has area an integer or half-integer number of square units. As an equilateral triangle of side-length d has area $A = \frac{\sqrt{3}}{4}d^2$, no lattice equilateral triangle can exist. (If one did, $\sqrt{3}$ would be rational.)

Challenge. A mathematician is an idealist! We've assumed that each dot on the square lattice is an ideal point of no width. Suppose that each point is actually a small disc of radius ε centered about a location with integer coordinates. Show that now it is possible to draw an equilateral triangle on the grid with each vertex lying on a dot of the grid. (See [ROUMAS and TANTON] for this and other dot-lattice drawing challenges.)

Equilateral polygons with any even number of sides exist. For example, here is an equilateral octagon. (It is clear how to extend this shape to produce an equilateral figure with a larger number of sides).

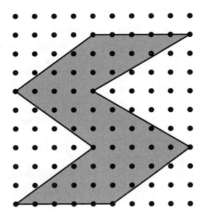

Curiously, it is impossible to construct an equilateral lattice figure with an odd number of sides. (It is not just an equilateral triangle that is impossible!)

We'll establish this in five steps.

Theorem. *No equilateral lattice figure with an odd number of sides exists.*

Proof. Suppose we had such a figure. Call its side-length d.

By Pythagoras's theorem we have that $d^2 = a^2 + b^2$ for two integers a and b. Because an even number squared is a multiple of four and an odd number squared is one more than a multiple of four, d^2 has remainder 0, 1, or 2 on division by four.

Let's see what must be mathematically wrong with this set-up.

Step One:

Draw a set of coordinate axes on the lattice so that the points of the lattice have integer coordinates. Starting at one vertex of the equilateral polygon, march around the figure taking note of the x-coordinate of each vertex. As there is an odd number of vertices, they cannot alternate in parity: there must be at least two neighboring vertices with x-coordinates both even or both odd. This means that there is a side of length d of the polygon with horizontal leg a even:

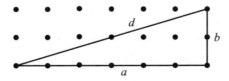

Thus d^2 can be written as the sum of two squares with at least one square being even. It follows that d^2 is congruent to 0 or 1 (but not 2) modulo four.

Step Two:

Call a vertex of the polygon *consistent* if its x- and y-coordinates are either both even or both odd, and call it *inconsistent* if its coordinates have opposite parity. As we march about the figure, there must be at least two neighboring vertices that are either both consistent or both inconsistent. This means it is possible to write d^2 either in the form $d^2 = (\text{even})^2 + (\text{even})^2$ or in the form $d^2 = (\text{odd})^2 + (\text{odd})^2$. The first forces d^2 to be congruent to 0 modulo four and the second congruent to 2 modulo four, which is not possible. Thus d^2 must be a multiple of four.

Step Three:

That d^2 is a multiple of four means that d^2 is of the form $(\text{even})^2 + (\text{even})^2$ for every side of the polygon. Thus all the coordinates of the vertices of the polygon have the same parity: all even or all odd. If the origin $(0, 0)$ is one of the vertices, then all vertices have even coordinates.

Step Four:

Now we are in trouble! If we take our odd-sided equilateral polygon and halve the values of the coordinates of the vertices, we obtain a smaller lattice polygon with the same odd number of sides and with the origin as one of the vertices. Moreover, the new polygon has one quarter the area of the original. Applying steps one, two, and three to the smaller polygon shows that it too has vertices with even coordinates, and we can then shrink it by a factor of two to obtain an even smaller lattice polygon with an odd number of sides. We could repeat this process to obtain a lattice polygon smaller than the unit square of the lattice! Our assumption that an odd-sided equilateral lattice polygon exists must be false.

$\square$

Challenge. A polygon is called *equiangular* if all of its interior angles have the same measure. A lattice rectangle is an example of a 4-sided equiangular lattice figure. As we have seen, no 3-sided equiangular lattice figure is possible. It is not too difficult to construct an example of an equiangular lattice octagon. Prove that if a lattice polygon is equiangular then it must have either 4 or 8 sides. (See [HONSBERGER, Section 13].)

Pick's Theorem

In the fall of 2009, students of the St. Mark's Institute of Mathematics research class studied Pick's theorem and discovered a beautiful and novel way to think about, prove, and generalize the theorem – an approach that is superior to the method of proof suggested in the newsletter. A short version of an article explaining their approach appeared in FOCUS, the news magazine of the Mathematical Association of America ([FOCUS]). The full version appears in Appendix II.

References

[FOCUS] Pick's Theorem – and Beyond! co-authored with St. Mark's Institute of Mathematics students, *FOCUS*, **30** no. 1 (Feb/March 2010), 14–35.

[HONSBERGER] Honsberger, R., *Mathematical Diamonds*, Mathematical Association of America, Washington D.C., (2003).

[ROUMAS and TANTON] Roumas, N. and Tanton, J., Lattice polygons for mathematicians and for engineers, *College Mathematics Journal*, **40**, no. 5, (2009), 336, 360, 369, 375.

[TANTON] Tanton, J., *Solve This: Math Activities for Students and Clubs*, Mathematical Association of America, Washington D.C., (2001).

13
Layered Tilings

PUZZLER: Structurally Sound Tilings

Here are two tilings of a 6 × 6 grid of squares using 1 × 2 tiles ("dominos").

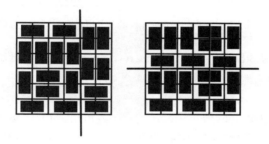

Each tiling is structurally flawed in the sense that each possesses a line along which it could slide.

Devise a tiling of a 6 × 6 grid that is structurally sound.

Is there a structurally sound tiling of an 8 × 8 board?

TIDBIT and RESEARCH: Layered Tilings

The following 2 × 3 board is double tiled with six dominos, meaning that each cell of the grid is covered by exactly two dominos.

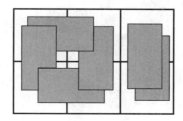

We'll say that a 2×3 grid is 2-tilable with dominos. It is also 3-tilable, 4-tilable and, in fact, k-tilable for any positive integer k.

Exercise. Draw an example of a 3-tiling for the 2×3 region.

Any region of cells that is 1-tilable with dominos (that is, can be tiled in the ordinary way) is also 2-tilable: layer an ordinary tiling on top of itself. Is the converse true? If a region is 2-tilable, is it also 1-tilable?

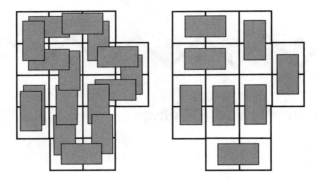

This 2-tilable region is also 1-tilable

Is it possible to create a region of squares that can be double layered with dominos but not single layered?

The Answer. Every double tiling of a region seems to divide it into loops. For instance, there are two loops in the picture above. This will always be the case. To see why, pick one half of any domino. It lies in a cell of the grid. The second half leads to a neighboring cell and a second domino. The second domino is connected to another cell and to another domino. In this way, the dominos trace out a path of connected cells. As there is only a finite number of cells in the grid, the loop must return to a previously visited cell (in fact, the starting cell) and so complete a loop. Repeat this process with another domino that wasn't visited to see that the entire grid of cells decomposes into disjoint loops of cells.

Having noticed this, it is now easy to convert a 2-tiling of a region into a 1-tiling by highlighting alternate dominos in each loop:

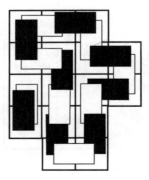

This shows that every 2-tiling is two 1-tilings layered on top of one another. In particular we have:

Any region 2-tilable with dominos is also 1-tilable

Next question: Must a region of cells that is 3-tilable also be 1-tilable? Or is it possible to design a region of cells that can be covered with a triple layer of dominos but not with a single layer?

This is a good question. Students of the St. Mark's Institute of Math research class did, and came up with an interesting answer!

Not all tiles are dominos. Consider the bent tromino, composed of three squares glued together to form an L-shape

It is possible for a region to be 3-tilable with bent trominos, but not be 1-tilable. The 2×2 grid is an example:

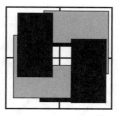

A 3-tilable region
that is <u>not</u> 1-tilable

Exercise. Find a region that is 2-tilable with bent trominos but is not 1-tilable.

Research. For every $k \geq 2$ is there an example of a region that is k-tilable with bent trominos but not 1-tilable?

We have two tile behaviors.
Dominos have the property that any 2-tilable region is actually 1-tilable. Bent trominos don't.

Exercise. Consider the 2×2 square tetromino composed of four unit squares

Prove that, like the domino, it has the same layer redundancy property: Any region that can be 2-tiled with square tetrominos can be 1-tiled.

Research. The straight tromino consists of three unit squares in a row

Does it have the layer redundancy property?

Research. Suppose a tile has the layer redundancy property, that any 2-tilable region is 1-tilable. Does it necessarily satisfy the stronger property: Any k-tilable region is 1-tilable? (Here k is a positive integer.)

Research. What can one say about layer redundancy for tiles on other grids? For example, the analog of a domino on a triangle grid is the lozenge.

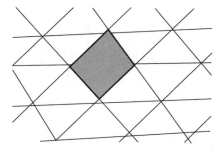

Is it layer redundant? Other tiles?

COMMENTARY, SOLUTIONS and THOUGHTS

The opening puzzler is a classic. (See [HONSBERGER], page 77). It is impossible to devise a tiling of a 6×6 grid that is structurally sound.

> In a 6×6 grid there are five horizontal and five vertical lines that each could be a line of slippage. Thus, in a sound tiling, we need to ensure that each is crossed by at least one domino.

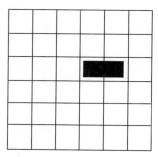

Actually, one won't suffice: one domino leaves an odd number of squares to be tiled on each side of the line, which cannot be done. Thus each potential line of slippage must be crossed by at least two dominos, requiring a minimum of 20 dominos. But for a 6 × 6 grid, we have only 18 dominos at our disposal. It is therefore impossible to accomplish the task.

For an 8 × 8 grid it can be done:

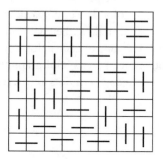

Layered Tilings

Students of the Fall 2007 St. Mark's Institute of Mathematics research class studied layered tilings. It was a challenging topic once we moved beyond dominos!

Here is a summary of the work of the group.

Recapitulating the Jargon

As we saw above, it is possible to tile a 2 × 3 grid of squares with dominos so that each cell is covered by exactly two tiles. We'll say that a 2 × 3 grid is (domino) 2-*tilable*.

Any region that is 1-tilable (that is, can be tiled with dominos in the ordinary sense) is k-tilable for any value k: stack dominos on top of one another in a single tiling. The converse is true: Any region that can be k-tiled with dominos is 1-tilable. (We'll prove this in a moment.) Thus the domino is *layer redundant* (LR).

Not all tiles have the layer redundancy property. For example, the bent tromino 3-tiles a 2 × 2 grid, but the 2 × 2 grid is not 1-tilable.

Does the straight tromino satisfy LR?

More generally: Can we classify the tiles that satisfy LR?

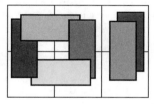

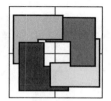

Dominos satisfy LR   Bent Trominos do not

Does the straight tromino satisfy LR?

Dominos Satisfy LR

As we saw in the newsletter, any region that can be 2-tiled with dominos is actually 1-tilable because each 2-tiling divides the region into disjoint loops so choosing every second domino from each loop leaves a single tiling of the board.

To generalize, the cells of a square grid can be colored black and white in a checkerboard fashion and each domino covers one cell of each color. If we represent the black and white cells as dots of the matching color and tiles as edges between dots, then any k-tiling of a region of squares can be represented as diagram of two rows of dots, one of each color, connected with lines between dots of opposite color, k lines from each dot. (In the language of graph theory, each dot is called a vertex, each line an edge, and the particular arrangement of dots and vertices described is called a bipartite graph.)

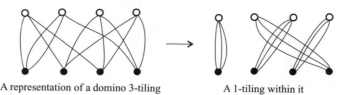

A representation of a domino 3-tiling   A 1-tiling within it

Again we can pursue the follow a loop idea. Start at a vertex and trace a path that, necessarily, returns to a previously visited vertex in an even number of steps. Consider the loop formed starting and ending at the vertex. Delete every second edge and draw a double edge for each of the remaining edges. This produces a new bipartite graph with each vertex of degree k with edges between vertices that were already connected by some edge. Repeat to transform the diagram into a match between the vertices, which represents a 1-tiling that was present in the original k-tiling. Thus the region is also 1-tilable.

This, by the way, proves the graph theoretic version of Hall's Matching Theorem. (See [CHARTRAND].)

If we remove the 1-tiling from the k-tiling we are left with a $(k-1)$-tiling of the region on which we can perform the same process. This shows that every k-tiling is the superposition of k 1-tilings, a stronger result. We have dominos completely licked!

Aside on Semi-Magic Squares

We can represent a tiling with dominos using an *adjacency array*. Given a k-tiling of a region of $2n$ cells, number the black cells 1 through n, in some order, and the white cells 1 through n. (Why must the number of cells be even?) Then construct an $n \times n$ matrix whose (i, j)th entry is the number of dominos lying across black cell i and white cell j.

$$\longrightarrow \quad \begin{pmatrix} 1 & 1 & 0 \\ 1 & 1 & 0 \\ 0 & 0 & 2 \end{pmatrix} = \begin{pmatrix} 1 & 0 & 0 \\ 0 & 1 & 0 \\ 0 & 0 & 1 \end{pmatrix} + \begin{pmatrix} 0 & 1 & 0 \\ 1 & 0 & 0 \\ 0 & 0 & 1 \end{pmatrix}$$

Every adjacency matrix that results is a semi-magic square, that is, an array of non-negative integers with the property that the entries in each row and in each column have the same sum, namely k, if the array arises from a k-tiling.

A 1-tiling corresponds to a semi-magic square with entries 0 and 1, with precisely one 1 appearing in each row and in each column. Such a matrix is called a *permutation matrix*.

The proof that dominos satisfy LR also establishes

Theorem. *Every semi-magic square is a sum of permutation matrices.*

Challenge. Suppose A and B are two $n \times n$ semi-magic squares. Prove that their sum $A + B$, their product $A \cdot B$ and the inverse A^{-1} (if it exists) are each semi-magic! (If J is the $n \times n$ matrix with all entries 1, then A is semi-magic with magic sum k if, and only if, $AJ = JA = kI$ where I is the identity matrix.)

Other Tiles

The student research group had very little success analyzing the layer redundancy property of other tiles. Here are a few observations.

1. *There is a region that is 2-tilable with bent trominos, but not 1-tilable.*

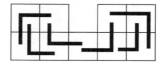

(For each $k > 1$, is there a region k-tilable with bent trominos but not 1-tilable?)

2. *The 2×2 square tetromino satisfies LR.*

Look at the top left cell of any region that is k-tiled with square tetrominos. It must be covered by k copies of the tile stacked on top of each other. Remove the stack and now analyse the k-tiling of the remaining region in the same way.

3. *A semi-magic cube need not be the sum of permutation cubes.*

This observation is relevant to the analysis of the straight tromino.

Analogous to the domino, there is a natural 3-coloring of the cells of a square grid so that no matter how the tile is placed on the grid, one cell of each color is covered.

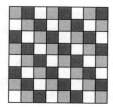

If we number the cells of each color 1 through n, we can associate with each tiling with straight trominos an adjacency cube. If we could prove that each semi-magic cube that arises is a sum of permutation cubes we'll have established that every k-tiling is a superposition of 1-tilings.

The $3 \times 3 \times 3$ cube with front face $\begin{pmatrix} 1 & 0 & 0 \\ 0 & 1 & 0 \\ 0 & 0 & 0 \end{pmatrix}$, the middle face $\begin{pmatrix} 0 & 0 & 1 \\ 0 & 0 & 0 \\ 0 & 1 & 0 \end{pmatrix}$ and back face $\begin{pmatrix} 0 & 0 & 0 \\ 1 & 0 & 0 \\ 0 & 0 & 1 \end{pmatrix}$ and magic sum 2 is semi-magic, but it is not the sum of permutation cubes. Hopes dashed? Not quite. Students were able to show that it cannot correspond to a 2-tiling of a square lattice region.

Challenge. Show that the ring octomino satisfies LR. For what it is worth, show that there is no possible coloring scheme of the cells of the square grid with eight colors with the property that no matter where an octomino is placed, one cell of each color is covered.

Challenge. Find an example of a region that is 2-tilable with straight trominos that possesses no 1-tiling within it. (Students found an example of such a region, but the region itself is 1-tilable!)

What is the key observation that brings the full story of layered tilings to light? It eluded our research group.

References

[CHARTRAND] Chartrand, G., *Introduction to Graph Theory*, Dover, New York, NY, 1985.

[HONSBERGER] Honsberger, R., *Mathematical Gems:From Elementary Combinatorics, Number Theory, and Geometry*, Mathematical Association of America, Washington D.C., 1973.

14

The Middle of a Triangle

PUZZLER

Most everyone would agree that the point of intersection of the two diagonals of a rectangle deserves to be called the "middle" of the rectangle. After all, it would the balance point of the shape if the rectangle were cut from uniform material and placed horizontally on the tip of a pencil. So maybe "middle" means "balance point."

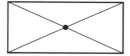

Challenge. Three squares are adjoined to make an L-shape piece. Find the location of the balance point of this figure. Does it seem reasonable to call this the "middle" of the L-shape?

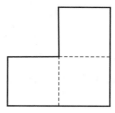

Design an L-shaped piece whose balance point lies <u>outside</u> the figure!

BONUS: Cake Puzzler

A rectangular piece is cut out from the interior of a rectangular cake. The edges of the piece taken are not necessarily aligned with the sides of the cake. Prove that with a single straight line cut one can divide what remains of the cake into two pieces of equal area. Moreover, explain where to make the cut.

TIDBIT: The Middle of a Triangle

Rectangles have "obvious" centers and most people would say, on first impulse, that every triangle probably has a well-defined center as well. Alas, the situation is not clear. There are many different candidates for the "middle" of a triangle. Here are five!

The Circumcenter. The *circumcenter O* of a triangle is the center of the circle that circumscribes the triangle.

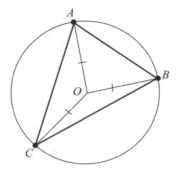

To see that such a circle exists from Pythagoras's theorem the points equidistant from two given points A and B are the points on the perpendicular bisector of $\overline{AB}$.

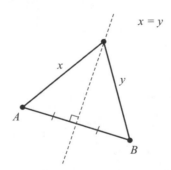

Suppose points A, B, and C are the vertices of a triangle. The perpendicular bisectors of two sides, $\overline{AB}$ and $\overline{BC}$, meet at some point O.

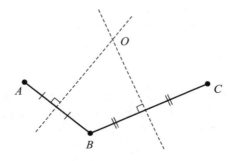

This point is simultaneously equidistant from A and B, and from B and C so it is equidistant from all three points! It is thus the center of a circle passing through A, B and C. (This also shows that the three perpendicular bisectors of a triangle are concurrent.)

The Incenter. The center I of a circle tangent to all three sides is called the *incenter* of the triangle.

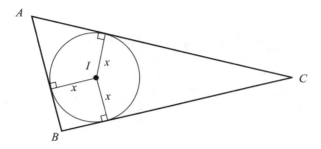

It is a point equidistant from all three sides.

To see that such a point exists, the angle bisector of two intersecting lines is a line equidistant from the two lines. (Use similar triangles to prove this.)

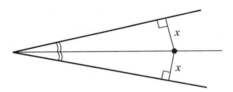

For $\triangle ABC$ above, the angle bisectors at A and at B meet at some point I and this point is simultaneously equidistant from all three sides of the triangle.

The Centroid. If the vertices of a triangle have coordinates $A = (x_1, y_1)$, $B = (x_2, y_2)$, $C = (x_3, y_3)$, then algebra shows that the point $G = \left(\frac{x_1+x_2+x_3}{3}, \frac{y_1+y_2+y_3}{3}\right)$ lies on each of the three medians of the triangle. (A median connects one vertex of the triangle to the midpoint of its opposite side.) This point is called the *centroid* of the triangle.

The Orthocenter. Suppose the triangle is positioned in the plane so that its circumcenter is at the origin. Then a (sneaky) exercise in algebra shows that the point $H = (x_1 + x_2 + x_3, \ y_1 + y_2 + y_3)$ lies on each of the three altitudes of the triangle. This point is the *orthocenter* of the triangle. As a bonus, we see now that O, G and H are collinear with G one third of the way along the segment $\overline{OH}$. This line on which they lie is call the *Euler line* of the triangle.

The Fermat Point. The point F inside the triangle that gives the smallest sum $FA + FB + FC$ is called the *Fermat point* of the triangle. A tough exercise in geometry shows that $\overline{FA}$, $\overline{FB}$, $\overline{FC}$ define three $120°$ angles. Also, if one draws equilateral triangles on the sides of $\triangle ABC$, then F lies on each of the lines connecting a vertex of $\triangle ABC$ to the far vertex of its opposite equilateral triangle.

Question. Which of these five centers is the "balance point" of the triangle?

Research Corner. Is there a triangle for which A, B, C, O, I, G, H, and F all have integer coordinates? (Points I and F are tough!)

COMMENTARY, SOLUTIONS and THOUGHTS

If two masses are connected by a straight rod of negligible weight, then the balance point of the system is somewhere on that straight line connecting the two. If they are of equal weight, then the balance point is at the midpoint of the rod. If not, then the balance point is shifted closer to the heavier mass, but is still on the line.

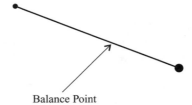

Balance Point

Without knowing any more than this, we can go quite far determining the location of balance points of geometrical figures.

A rectangle of mass M made of uniformly dense material is equivalent, in terms of its physics, to point mass M situated at its center (where the diagonals intersect). Thus the balance point of a system of two rectangles is a point somewhere on the line connecting their centers.

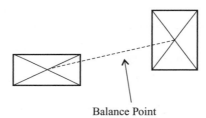

Balance Point

To answer the puzzle, an L-shaped piece can be seen as the union of two rectangles in two different ways. In each, the balance point of the figure is along the line connecting the rectangle centers. As there are two such lines, the balance point must be at their point of intersection!

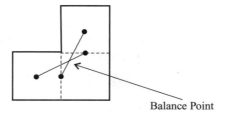

Balance Point

It is not hard to see that if the legs of the L-shape are sufficiently long, then the balance point of the figure lies outside of the figure.

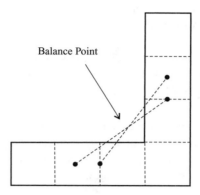

Comment. One can regard each of these L-shaped figures as the union of two squares, one composed of material of positive density and one composed of material of matching negative density that cancels positive mass. (Imagine both squares with top-left corners aligned. Only the L-shaped piece remains visible.) This is the kind of thinking that solves the cake puzzler:

> *The straight cut that passes through the center of the missing rectangular piece and the center of the entire cake cuts what remains of the cake into two pieces of equal area.*

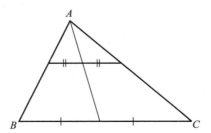

The same solution applies to a circular cake with a regular hexagonal hole and a regular decagonal shape with a square hole, and so on. (Are regular polygons, rectangles, and circles the only shapes that possess an interior point with the property that any line through it cuts its area exactly in half?)

The Balance Point of a Triangle

Similarity shows that the median of a triangle from its apex A to base $\overline{BC}$ divides each line segment within the triangle that is parallel to the base into two equal parts.

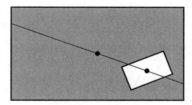

Suppose the triangle is made of a uniformly dense material. If we approximate it as a series of stacked rectangular strips with center points collinear along the median, then that system of strips would balance if placed on a knife's edge following the median. Taking finer and finer strips we can argue then that the triangle itself would balance along the median. The balance point of the triangle must therefore lie on this line.

There is nothing special here about the median from vertex A, so the balance point must lie on the other two medians as well. Thus the balance point is the point where the three medians meet.

The balance point of a triangle is its centroid.

Comment. Suppose three unit masses are placed at three locations A, B and C. Then the balance point of the three-mass system lies someone on the line connecting point A with the midpoint of $\overline{BC}$ (the balance point of the two masses B and C). The balance point of a three-mass system is thus the centroid of the triangle formed by those three masses.

The Coordinates of Central Points

In the newsletter we claimed that if the three vertices of a triangle are $A = (x_1, y_1)$, $B = (x_2, y_2)$, and $C = (x_3, y_3)$, then the centroid of the triangle has coordinates $G = \frac{1}{3}(x_1 + x_2 + x_3, y_1 + y_2 + y_3)$ and, further, if the circumcenter of the triangle lies at the origin, then the orthocenter has coordinates $H = (x_1 + x_2 + x_3, y_1 + y_2 + y_3)$. Let's prove these assertions.

If $P = (p_1, p_2)$ and $Q = (q_1, q_2)$ are two points in the plane, then the vector $\overrightarrow{PQ}$ that shifts P to Q is given by

$$\overrightarrow{PQ} = \langle q_1 - p_1, q_2 - p_2 \rangle = \text{``}Q - P\text{''}$$

(so that $P + \overrightarrow{PQ} = P + (Q - P)$ is indeed Q).

The midpoint M of $\overline{BC}$ is given by

$$M = B + \frac{1}{2}\overrightarrow{BC} = \left(x_2 + \frac{1}{2}(x_3 - x_2), y_2 + \frac{1}{2}(y_3 - y_2) \right) = \frac{1}{2}(x_2 + x_3, y_2 + y_3)$$

and the point one-third of the way along from M to A is

$$\begin{aligned}
M + \frac{1}{3}\overrightarrow{MA} &= \frac{2}{3}M + \frac{1}{3}A \\
&= \frac{1}{3}(x_2 + x_3, y_2 + y_3) + \frac{1}{3}(x_1, y_1) \\
&= \frac{1}{3}(x_1 + x_2 + x_3, y_1 + y_2 + y_3).
\end{aligned}$$

This is symmetrical among the coordinates and so must also be the point one-third of the way along each of the other two medians as well. It is a common point to all three medians and so must be G, the centroid of the triangle.

Suppose now that we have placed $\triangle ABC$ so that $A = (x_1, y_1)$, $B = (x_2, y_2)$, and $C = (x_3, y_3)$ are the same distance from the origin. Consider the point $P = (x_1 + x_2 + x_3, \ y_1 + y_2 + y_3)$.

Then

$$\overrightarrow{AP} = P - A = (x_2 + x_3, \ y_2 + y_3),$$
$$\overrightarrow{BC} = B - C = (x_2 - x_3, \ y_2 - y_3)$$

and

$$\overrightarrow{AP} \cdot \overrightarrow{BC} = x_2^2 - x_3^2 + y_2^2 - y_3^2 = \left(x_2^2 + y_2^2\right) - \left(x_3^2 + y_3^2\right) = 0,$$

and so the two vectors are perpendicular. This shows that P lies on the altitude from A. Similar algebra shows that P also lies on the remaining two altitudes and so P is, in fact, the orthocenter H of the circle.

Challenge. If the vertices of a triangle are $A = (x_1, y_1)$, $B = (x_2, y_2)$, and $C = (x_3, y_3)$, show that the incenter of the triangle has coordinates

$$I = \left(\frac{ax_1 + bx_2 + cx_3}{a + b + c}, \ \frac{ay_1 + by_2 + cy_3}{a + b + c}\right)$$

where a, b, and c are the lengths of the sides opposite A, B, and C respectively.

The Fermat Point

If one placed three small pegs between two plastic sheets, each peg perpendicular to the sheets, and dipped the model into a soap solution, one of two things might occur: three rectangular surfaces of soap film will form outlining the triangle defined by the three pegs, or three rectangular surfaces will form meeting at a point somewhere in the interior of the triangle defined by the pegs. A soap film "pulls in on itself" and wants to minimize surface area, so in the second case the interior point where the film meets will be the Fermat point of the triangle.

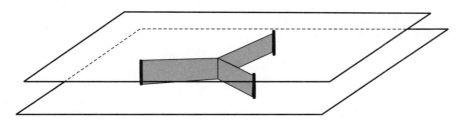

Proving that the Fermat point exists for any triangle is somewhat delicate. [HONSBERGER, pages 24–34] and [COURANT and ROBBINS, pages 354–359] show details. (An internet search also yields derivations.) If the triangle has an interior angle greater than or equal to $120°$, then the Fermat point of the triangle is the vertex of the triangle at that obtuse angle.

For a list of other possible candidates for the "middle" of a triangle, see [KIMBERLING].

References

[COURANT and ROBBINS] Courant, R. and Robbins, H., *What Is Mathematics?, 2nd ed.*, Oxford University Press, Oxford, England, 1941.

[HONSBERGER] Honsberger, R., *Mathematical Gems I,* Mathematical Association of America, Washington D. C., 1973.

[KIMBERLING] Kimberling, C., "Triangle centers." URL: http://faculty.evansville.edu/ck6/tcenters/index.html

15
Partitions

PUZZLER: Partitions

A *partition* of a counting number N is an expression that represents N as a sum of counting numbers. For example, there are eight partitions of 4 if order is considered important:

$$4 \quad 3+1 \quad 1+3 \quad 2+2$$
$$2+1+1 \quad 1+2+1 \quad 1+1+2 \quad 1+1+1+1$$

There are five unordered partitions of 4:

$$4 \quad 3+1 \quad 2+2 \quad 2+1+1 \quad 1+1+1+1$$

a) How many ordered partitions are there of the numbers 1 through 6. Any patterns?

b) How many unordered partitions are there of the numbers 1 through 6. Any patterns? Will these partition numbers continue to be prime?

   One can place restrictions on the types of partitions one wishes to count. For example, there are eight partitions of 10 with exactly three terms, order not important:

$$8+1+1 \quad 7+2+1 \quad 6+3+1 \quad 6+2+2$$
$$5+4+1 \quad 5+3+2 \quad 4+4+2 \quad 4+3+3$$

c) Show that there are eight ways to partition 10 using only 1, 2, and 3, order immaterial, using at least one 3.

d) Let $A(N)$ be the number of ways to partition N as a sum of exactly three terms, and let $B(N)$ be the number of partitions of N with largest term 3. (Order immaterial in both cases.) Part c) shows $A(10) = 8 = B(10)$. Prove that $A(N) = B(N)$ always.

TIDBIT

A partition of a number can be represented as columns of dots. For example, the partition $6 + 3 + 1$ of 10 can be depicted as

If order is not considered important, then it is customary to place the terms in order from biggest to smallest. From now on we shall assume that for all our partitions order is indeed immaterial. If one looks at this diagram sideways, we see the alternative partition $3 + 2 + 2 + 1 + 1 + 1$ of 10. The two partitions are said to be "conjugate" partitions. Some partitions are "self-conjugate." For example, the partition $8 + 6 + 4 + 3 + 2 + 2 + 1 + 1$ of the number 27 is self-conjugate.

The diagram of any self-conjugate partition can seen as a series of nested L-shapes, each L with an odd number of dots.

Also, no two L shapes can contain the same number of odd dots.

This shows that each self-conjugate partition of a number gives rise to a partition into distinct odd terms, as 27 can be written $15 + 9 + 3$. Conversely, any partition into distinct odd terms can be seen as arising from a self-conjugate partition. So:

The count of self-conjugate partitions of a number equals the count of partitions into distinct odd numbers.

One of Euler's Remarkable Results

In 1740 French mathematician Philippe Naudé sent a letter to Leonhard Euler asking him about the number of ways a positive integer could be written as a sum of distinct positive integers (order immaterial). Euler examined the problem and discovered something truly remarkable.

Let $D(n)$ be the number of ways to write n as a sum of distinct positive integers. For instance, $D(6) = 4$ because we can write six as 6, $5 + 1$, $4 + 2$, or $3 + 2 + 1$.

Exercise. Find $D(2)$, $D(5)$, and $D(13)$.

Let $O(n)$ be the number of ways to write n as a sum of odd integers. For example, $O(6) = 4$ because we can write six as $5 + 1$ or $3 + 3$ or $3 + 1 + 1 + 1$ or $1 + 1 + 1 + 1 + 1 + 1$.

Exercise. Find $O(2)$, $O(5)$ and $O(13)$.

EULER proved that $O(N) = D(N)$ **always.**

Euler's Proof. What would happen if one expanded the infinite quantity

$$(1 + x)(1 + x^2)(1 + x^3)(1 + x^4)\cdots?$$

By selecting "1" from each set of parentheses we would obtain the constant term 1. By selecting x and then 1s thereafter, the term x will arise. The term x^2 will arise by selecting 1, then x^2 and then 1s thereafter. Skipping ahead, how will the term x^6 arise? It arises in more than one way: As x^6 and 1s elsewhere, as x and x^5 and 1s elsewhere, as x^2 and x^4 and 1s, as x^3 and x^2 and x and 1s. There are four ways x^6 appears and they match the partitions of 6 into distinct numbers: $6, 5 + 1$, $4 + 2, 3 + 2 + 1$. In general, the coefficient of x^N in the expansion of the infinite product is $D(N)$. We have:

$$(1 + x)(1 + x^2)(1 + x^3)(1 + x^4)\cdots$$
$$= 1 + x + x^2 + 2x^3 + 2x^4 + 3x^5 + 4x^6 + \cdots$$

Now consider $\frac{1}{1-x} \cdot \frac{1}{1-x^3} \cdot \frac{1}{1-x^5} \cdot \frac{1}{1-x^7} \cdots$. Because

$$\frac{1}{1 - z} = 1 + z + z^2 + z^3 + \cdots$$

this infinite product is

$$(1 + x + x^{1+1} + x^{1+1+1} + \cdots)$$
$$\times (1 + x^3 + x^{3+3} + x^{3+3+3} + \cdots)$$
$$\times (1 + x^5 + x^{5+5} + x^{5+5+5} + \cdots)$$
$$\times \cdots$$

What do we obtain if we expand this? By selecting 1 from each set of parentheses there is a constant term of 1. There will also be a single x term, and a single x^2 term (from selecting x^{1+1}). There are four ways x^6 arises: as x^{3+3}, as $x^5 \cdot x^1$, as $x^3 \cdot x^{1+1+1}$, and as $x^{1+1+1+1+1+1}$. We see that x^N arises from the odd partitions of N. The coefficient of x^N is $O(N)$.

If we can prove that the two infinite products are equal, then it follows that their expansions are equal. This will give $O(N) = D(N)$.

Okay. Here goes!

$$\frac{1}{1 - x} \cdot \frac{1}{1 - x^3} \cdot \frac{1}{1 - x^5} \cdot \frac{1}{1 - x^7} \cdots$$
$$= \frac{1}{1 - x} \cdot \frac{1 - x^2}{1 - x^2} \cdot \frac{1}{1 - x^3} \cdot \frac{1 - x^4}{1 - x^4} \cdot \frac{1}{1 - x^5} \cdot \frac{1 - x^6}{1 - x^6} \cdot \frac{1}{1 - x^7} \cdots$$
$$= \frac{1}{1 - x} \cdot \frac{(1 - x)(1 + x)}{1 - x^2} \cdot \frac{1}{1 - x^3} \cdot \frac{(1 - x^2)(1 + x^2)}{1 - x^4} \cdot \frac{1}{1 - x^5} \cdot \frac{(1 - x^3)(1 + x^3)}{1 - x^6} \cdot \frac{1}{1 - x^7} \cdots$$

Cross out common terms and we are left with: $(1 + x)(1 + x^2)(1 + x^3)(1 + x^4)\cdots$. DONE!

Euler was not at all bashful about playing with the infinite!

Research Corner. Let $P(N)$ be the number of unordered partitions of N with no restrictions. No one knows a formula for $P(N)$. Might you find one?

COMMENTARY, SOLUTIONS and THOUGHTS

The theory of partitions is a vast topic still open to much further research and study.

Let $P(N)$ be the count of unordered partitions of N subject to no restrictions. (These numbers are often called the partition numbers.) Their first ten values are:

N	1	2	3	4	5	6	7	8	9	10
$P(N)$	1	2	3	5	7	11	15	22	30	42

and this sequence of numbers continues to grow rapidly without bound: $P(100) = 190,569,292$ and $P(1000) \approx 2.4 \times 10^{31}$. Indian mathematician Srinivasa Ramanujan guessed that $P(N)$ is well approximated as:

$$P(N) \approx \frac{1}{4\pi\sqrt{3}} e^{\pi\sqrt{\frac{2N}{3}}}$$

and was later able to prove this as valid with the help of G.H. Hardy in 1918 ([HARDY and RAMANUJAN]). In January 2011, Ken Ono and his collaborators announced their discovery of self-repeating patterns within the sequence of partition numbers. They believe this fractal-like structure will provide the means to find an exact formula for these numbers. (See [SALERNO].)

Ordered partitions, on the other hand, are straightforward to analyze. For example, there are eight ordered partitions of 4, and if we use dashes to represent numerals these can be expressed as

$$\begin{aligned}
4 &= |\ |\ |\ | & |+|+|\ | &= 1+1+2 \\
3+1 &= |\ |\ |+| & |+|\ |+| &= 1+2+1 \\
2+2 &= |\ |+|\ | & |\ |+|+| &= 2+1+1 \\
1+3 &= |+|\ |\ | & |+|+|+| &= 1+1+1+1
\end{aligned}$$

There are three spaces between the four dashes that can be left blank or filled with a plus sign and so there are 2^3 options, yielding eight ordered partitions. In general, a number N can be represented as N dashes with $N-1$ spaces, yielding 2^{N-1} ordered partitions.

Question. How many ordered partitions of the number 50 have exactly 17 terms?

Euler's Pentagonal Number Theorems

I cannot resist taking the ideas of this newsletter an extra step (well, several steps!) to describe Euler's most peculiar and astounding discovery in partition theory. One can learn more of this mathematics in [HARDY] and in [TATTERSALL], for instance. Robert Young in [YOUNG] presents a translated passage from Euler's memoir that describes Euler's process in discovering and establishing this piece of mathematics.

We saw in the newsletter that the infinite product

$$(1+x)(1+x^2)(1+x^3)(1+x^4)\cdots,$$

when expanded, generates the numbers $D(N)$, the count of ways to write N as a sum of distinct terms, order immaterial. ($D(N)$ is the coefficient of x^N.) We have

$$(1 + x)(1 + x^2)(1 + x^3)(1 + x^4) \cdots = 1 + x + x^2 + 2x^3 + 2x^4 + 3x^5 + 4x^6 + \cdots.$$

Euler wondered about the related infinite product

$$(1 - x)(1 - x^2)(1 - x^3)(1 - x^4) \cdots.$$

When expanded it is

$$(1 - x)(1 - x^2)(1 - x^3)(1 - x^4) \cdots$$
$$= 1 - x - x^2 + x^5 + x^7 - x^{12} - x^{15} + x^{22} + x^{26} - x^{35} - x^{40} + \cdots.$$

The signs seem to alternate in pairs and the exponents that appear are

$$0, \ 1, \ 2, \ 5, \ 7, \ 12, \ 15, \ 22, \ 26, \ 35, \ 40, \ldots.$$

Euler observed that taking the difference between consecutive terms yields the sequence

$$1, \underline{1}, 3, \underline{2}, 5, \underline{3}, 7, \underline{4}, 9, \underline{5}, 11, \ldots,$$

which is the sequence of counting numbers $1, 2, 3, 4, \ldots$ intertwined with the sequence of odd numbers $1, 3, 5, 7, 9, \ldots$. This allows us to generate additional terms easily. As Euler observed, we can use induction to verify these claims but that does not offer clarity on an intuitive level. How does $(1 - x)(1 - x^2)(1 - x^3)(1 - x^4) \cdots$ connect with partitions?

Getting a Handle on this Infinite Product

In expanding $(1 + x)(1 + x^2)(1 + x^3)(1 + x^4) \cdots$ each partition $a + b + c + \cdots + z$ of N into distinct terms contributes once to the coefficient of x^N (from selecting $+x^a$, $+x^b$, $+x^c$ and so on, and 1s elsewhere among the parentheses). In expanding $(1 - x)(1 - x^2)(1 - x^3)(1 - x^4) \cdots$ the partition $a + b + c + \cdots + z$ contributes $+1$ to the coefficient of x^N if it involves an even number of terms, and -1 if it involves an odd number of terms.

Let $D_{even}(N)$ and $D_{odd}(N)$ denote, respectively, the count of ways to write N as a sum of an even or odd number of distinct terms. We see

$$(1 - x)(1 - x^2)(1 - x^3)(1 - x^4) \cdots$$
$$= 1 + (D_{even}(1) - D_{odd}(1))x + (D_{even}(2) - D_{odd}(2))x^2 + \cdots$$
$$+ (D_{even}(N) - D_{odd}(N))x^N + \cdots.$$

In 1881 Fabian Franklin noted that a partition of a number into an even count of distinct parts often matches a partition of that number into an odd count of distinct parts. We'll deviate from Euler's approach at this point (and return to it later on) and explain the work of Franklin. It is relevant.

A diagram of a partition possesses a rightmost column of dots and, from the top left dot, a southwest diagonal of dots at a 45 degree slope. Let r denote the number of dots in its rightmost column and d the number of dots in the diagonal. (Figure 1(a) has $r = 2$ and $d = 3$.) If r is smaller than d, then moving the r dots in the rightmost column to lay them on top of the diagonal produces a partition still with distinct terms, but one fewer. If r is bigger than $d + 1$, then moving dots from the diagonal and making a new rightmost column also creates a new partition with distinct terms but now with one more term. In each case, the transformations change the parity of the partition.

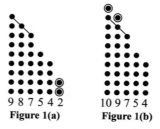

9 8 7 5 4 2   10 9 7 5 4
Figure 1(a)   **Figure 1(b)**

The first transformation also works for $r = d$ as long as the dot at the top of the rightmost column is not part of the diagonal. (Figure 2(a) illustrates this problem.) The second transformation also works for $r = d + 1$ as long as no dot lies both in the rightmost column and the diagonal, as in Figure 2(b).

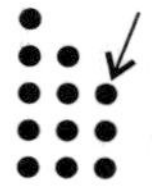

$r = d$ $r = d+1$
Figure 2(a) 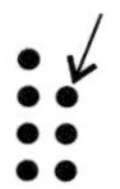   **Figure 2(b)**

If the first case occurs, then N is a number of the form

$$\frac{1}{2}d \times (d + (d - 1) + d) = \frac{d(3d - 1)}{2}$$

(the dots form half a rectangle) and there is one partition of N into d distinct terms that we cannot match with a different partition. All other partitions match in pairs.

If the second case occurs, then N is a number of the form

$$\frac{1}{2}d \times (d + (d + 1) + d) = \frac{d(3d + 1)}{2}$$

and there is this partition of N into d distinct terms that we cannot match with a different partition. All other partitions match in pairs.

For other values of N (that is, those not of the form $\frac{d(3d\pm1)}{2}$), even and odd partitions of N occur in pairs.

Thus we have

$$D_{even}(N) - D_{odd}(N) = 0 \quad \text{if } N \neq \frac{d(3d \pm 1)}{2}.$$

$$D_{even}(N) - D_{odd}(N) = 1 \quad \text{if } N = \frac{d(3d \pm 1)}{2} \text{ with } d \text{ even.}$$

$$D_{even}(N) - D_{odd}(N) = -1 \quad \text{if } N = \frac{d(3d \pm 1)}{2} \text{ with } d \text{ odd.}$$

Substituting $d = 1, 2, 3, 4, \ldots$ into $\frac{d(3d\pm1)}{2}$ gives the sequence $1, 2, 5, 7, 12, 15, \ldots$ with $d = 1$ giving 1 and 2 (d odd corresponds to a coefficient of $+1$), $d = 2$ giving the terms 5 and 7 (d even corresponds to a coefficient of -1), $d = 3$ giving the terms 12 and 15, and so on. From

$$(1 - x)(1 - x^2)(1 - x^3)(1 - x^4) \cdots$$
$$= 1 + (D_{even}(1) - D_{odd}(1))x + (D_{even}(2) - D_{odd}(2))x^2 + (D_{even}(3) - D_{odd}(3))x^3 + \cdots$$

we now see that

$$(1 - x)(1 - x^2)(1 - x^3)(1 - x^4) \cdots$$
$$= 1 + (-1)x + (-1)x^2 + (0)x^3 + (0)x^4 + (+1)x^5 + (0)x^6 + (+1)x^7 + \cdots$$
$$= 1 - x - x^2 + x^5 + x^7 - \cdots,$$

explaining the alternating pairs of signs and the numbers appearing as exponents.

Connection to General Partitions

We have seen that $(1 + x)(1 + x^2)(1 + x^3)(1 + x^4) \cdots$ generates partitions in which each number is used at most once (the partitions into distinct terms).

The product $(1 + x^1 + x^{1+1})(1 + x^2 + x^{2+2})(1 + x^3 + x^{3+3})(1 + x^4 + x^{4+4}) \cdots$ generates partitions that use each number at most twice, the product $(1 + x^1 + x^{1+1} + x^{1+1+1})(1 + x^2 + x^{2+2} + x^{2+2+2})(1 + x^3 + x^{3+3} + x^{3+3+3})(1 + x^4 + x^{4+4} + x^{4+4+4}) \cdots$ those that use each number at most three times, and so on.

The infinite product

$$(1 + x^1 + x^{1+1} + x^{1+1+1} + x^{1+1+1+1} + \cdots)$$
$$\times (1 + x^2 + x^{2+2} + x^{2+2+2} + x^{2+2+2+2} + \cdots)$$
$$\times (1 + x^3 + x^{3+3} + x^{3+3+3} + x^{3+3+3+3} + \cdots)$$
$$\times (1 + x^4 + x^{4+4} + x^{4+4+4} + x^{4+4+4+4} + \cdots)$$
$$\times \cdots$$

generates the partitions with no limit on the number of times each integer is used. That is, it generates the numbers $P(N)$. [Check this by showing that there are seven ways to produce the term x^5, each corresponding to a partition of 5.]

We have then that this infinite product equals

$$1 + P(1)x + P(2)x^2 + P(3)x^3 + P(4)x^4 + \cdots.$$

The geometric series formula shows that the infinite product also equals

$$\frac{1}{1 - x} \cdot \frac{1}{1 - x^2} \cdot \frac{1}{1 - x^3} \cdots,$$

which tells us that multiplying it by $(1 - x)(1 - x^2)(1 - x^3)(1 - x^4) \cdots$ yields the value 1. That is, multiplying

$$1 + P(1)x + P(2)x^2 + P(3)x^3 + P(4)x^4 + \cdots$$

by

$$1 - x - x^2 + x^5 + x^7 - x^{12} - x^{15} + x^{22} + x^{26} - x^{35} - x^{40} + \cdots$$

gives 1. Let's write this out:

$$
\begin{aligned}
1 &= (1 + P(1)x + P(2)x^2 + P(3)x^3 + P(4)x^4 + \cdots)(1 - x - x^2 + x^5 + x^7 - x^{12} - x^{15} + \cdots) \\
&= 1 + P(1)x + P(2)x^2 + P(3)x^3 + P(4)x^4 + P(5)x^5 + P(6)x^6 + P(7)x^7 + \cdots \\
&\quad\ - x - P(1)x^2 - P(2)x^3 - P(3)x^4 - P(4)x^5 - P(5)x^6 - P(6)x^7 - \cdots \\
&\quad\ - x^2 - P(1)x^3 - P(2)x^4 - P(3)x^5 - P(4)x^6 - P(5)x^7 - \cdots \\
&\qquad\qquad\qquad\qquad\quad + x^5 + P(1)x^6 + P(2)x^7 + \cdots \\
&\qquad\qquad\qquad\qquad\qquad\qquad\qquad + x^7 + \cdots \\
&\qquad\qquad\qquad\qquad\qquad\qquad\qquad\qquad - \cdots
\end{aligned}
$$

Thus we can see

$$
\begin{aligned}
P(1) - 1 &= 0 \\
P(2) - P(1) - 1 &= 0 \\
P(3) - P(2) - P(1) &= 0 \\
P(4) - P(3) - P(2) &= 0 \\
P(5) - P(4) - P(3) + 1 &= 0 \\
P(6) - P(5) - P(4) + P(1) &= 0 \\
P(7) - P(6) - P(5) + P(2) &= 0.
\end{aligned}
$$

And, in general

$$
\boxed{P(N) - P(N-1) - P(N-2) + P(N-5) + P(N-7) - P(N-12) - P(N-15) + \cdots = 0}
$$

with the convention that $P(N - k)$ is zero if $N - k$ is negative and $P(N - k)$ is 1 if $N - k$ is zero.

Although we have no explicit formula for $P(N)$ we have established a recursion relation among the partition numbers fundamentally connected to the numbers $1, 2, 5, 7, 12, 15, \ldots$. This unexpected result is known as Euler's Pentagonal Number Theorem.

Comment. The numbers $1, 5, 12, 22, \ldots$ arising as every second term in the sequence are the *pentagonal numbers*. They come from arranging dots into pentagonal arrays as in Figure 3.

1   5   12   22

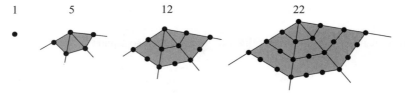

Figure 3

The k th pentagonal is composed of a large triangle and two smaller triangles. (Figure 4.)

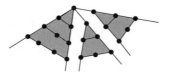

Figure 4

If $T_k = 1 + 2 + 3 + \cdots + k = \frac{k(k+1)}{2}$ denotes the kth triangular number, then the kth pentagonal number is $T_k + 2T_{k-1} = \frac{k(k+1)+2(k-1)k}{2} = \frac{k(3k-1)}{2}$.

Sums of Divisors

Euler played with the formula

$$(1-x)(1-x^2)(1-x^3)(1-x^4)\cdots$$
$$= 1 - x - x^2 + x^5 + x^7 - x^{12} - x^{15} + x^{22} + x^{26} - x^{35} - x^{40} + \cdots$$

a second way and discovered another pentagonal theorem in a different context.

Warning. Calculus!
The number 12 has factors 1, 2, 3, 4, 6, and 12 and their sum is $1 + 2 + 3 + 4 + 6 + 12 = 28$. For a positive integer N let $\sigma(N)$ denote the sum of its factors.

Comment. This quantity has been of interest since the time of the ancient Greeks: a number is dubbed "perfect" if all but the largest factor sum to the number itself. Thus a number is perfect if $\sigma(N) = 2N$. For example, 6 is perfect as it has factors 1, 2, 3, 6 and $1 + 2 + 3 = 6$. The first five perfect numbers are 6, 28, 496, 8128, and 33550336, all triangular and all even. Euler proved that all even perfect numbers must be triangular. See [TATTERSALL], for instance. No one knows an example of an odd perfect number nor has it been proved that none exist.

Let

$$S = (1-x)(1-x^2)(1-x^3)(1-x^4)\cdots \tag{1}$$

We know

$$S = 1 - x - x^2 + x^5 + x^7 - x^{12} - x^{15} + x^{22} + x^{26} - x^{35} - x^{40} + \cdots \tag{2}$$

Take the logarithm of S from (1)

$$\ln S = \ln(1-x) + \ln(1-x^2) + \ln(1-x^3) + \cdots$$

and differentiate

$$\frac{1}{S} \cdot \frac{dS}{dx} = -\frac{1}{1-x} - \frac{2x}{1-x^2} - \frac{3x^2}{1-x^3} - \frac{4x^3}{1-x^4} - \cdots$$

or

$$-\frac{x}{S} \cdot \frac{dS}{dx} = \frac{x}{1-x} + \frac{2x^2}{1-x^2} + \frac{3x^3}{1-x^3} + \frac{4x^4}{1-x^4} + \cdots.$$

Comment. Like Euler, we do not worry about whether we can take a logarithm of an infinite product or differentiate an infinite sum term by term. One should, of course, at some point fret about such details!

Using the geometric series formula we can express this as

$$-\frac{x}{S} \cdot \frac{dS}{dx} = x(1 + x + x^2 + x^3 + \cdots) + 2x^2(1 + x^2 + x^4 + \cdots)$$
$$+ 3x^3(1 + x^3 + x^6 + x^9 + \cdots) + \cdots$$

or

$$-\frac{x}{S}\cdot\frac{dS}{dx} = \begin{aligned} &x + x^2 + x^3 + x^4 + x^5 + x^6 + x^7 + x^8 + \cdots \\ & + 2x^2 \phantom{{}+ x^3} + 2x^4 \phantom{{}+ x^5} + 2x^6 \phantom{{}+ x^7} + 2x^8 + \cdots \\ & + 3x^3 \phantom{{}+ x^4 + x^5} + 3x^6 \phantom{{}+ x^7} + \cdots \\ & + 4x^4 \phantom{{}+ x^5 + x^6 + x^7} + 4x^8 + \cdots \\ & + 5x^5 \phantom{{}+ x^6 + x^7 + x^8} + \cdots \\ & + 6x^6 \phantom{{}+ x^7 + x^8} + \cdots \\ & + 7x^7 \phantom{{}+ x^8} + \cdots \\ & + 8x^8 + \cdots \\ & + \cdots \end{aligned}$$

Looking down the columns we see that factors of each exponent appear. For example, for fourth powers we see the terms x^4 and $2x^4$ and $4x^4$ and so the coefficient of x^4 is $1 + 2 + 4 = 7 = \sigma(4)$. In general,

$$-\frac{x}{S}\cdot\frac{dS}{dx} = \sigma(1)x + \sigma(2)x^2 + \sigma(3)x^3 + \sigma(4)x^4 + \cdots$$

From (2)

$$\frac{dS}{dx} = -1 - 2x + 5x^4 + 7x^6 - 12x^{11} - 15x^{15} + \cdots$$

and so

$$-\frac{x}{S}\cdot\frac{dS}{dx} = \frac{x + 2x^2 - 5x^5 - 7x^7 + 12x^{12} + 15x^{15} - \cdots}{1 - x - x^2 + x^5 + x^7 - x^{12} - x^{15} + \cdots}$$

We now have two expressions for $-\frac{x}{S}\cdot\frac{dS}{dx}$ so

$$\sigma(1)x + \sigma(2)x^2 + \sigma(3)x^3 + \sigma(4)x^4 + \cdots = \frac{x + 2x^2 - 5x^5 - 7x^7 + 12x^{12} + 15x^{15} - \cdots}{1 - x - x^2 + x^5 + x^7 - x^{12} - x^{15} + \cdots}.$$

Multiplying through by the denominator and subtracting yields

$$\begin{aligned} 0 = &\sigma(1)x + \sigma(2)x^2 + \sigma(3)x^3 + \sigma(4)x^4 + \sigma(5)x^5 + \sigma(6)x^6 + \cdots \\ &- x - \sigma(1)x^2 - \sigma(2)x^3 - \sigma(3)x^4 - \sigma(4)x^5 - \sigma(5)x^6 - \cdots \\ &- 2x^2 - \sigma(1)x^3 - \sigma(2)x^4 - \sigma(3)x^5 - \sigma(4)x^6 - \cdots \\ &+ 5x^5 + \sigma(1)x^6 + \cdots \\ &+ \cdots \end{aligned}$$

Thus

$$\begin{aligned} \sigma(1) &= 1 \\ \sigma(2) &= \sigma(1) + 2 \\ \sigma(3) &= \sigma(2) + \sigma(1) \\ \sigma(4) &= \sigma(3) + \sigma(2) \\ \sigma(5) &= \sigma(4) + \sigma(3) - 5 \\ \sigma(6) &= \sigma(5) + \sigma(4) - \sigma(1) \end{aligned}$$

and in general

$$\sigma(N) = \sigma(N-1) + \sigma(N-2) - \sigma(N-5) - \sigma(N-7) + \sigma(N-12) + \sigma(N-15) - \cdots$$

with the convention here that $\sigma(N-k)$ is zero if $N-k$ is negative, and $\sigma(N-k)$ is N if $N-k$ is zero.

This result is also known as Euler's Pentagonal Theorem.

References

[HARDY and RAMANUJAN] Hardy, G. H., and Ramanujan, S., Asymptotic formulae in combinatory analysis, *Proceedings of the London Mathematics Society*, **17** (1918), 75–115.

[HARDY and WRIGHT] Hardy, G. H., and Wright, E. M., *An Introduction to the Theory of Numbers*, Clarendon Press, Oxford, England,1979.

[SALERNO] Salerno, A., "New Theorems: Partition Numbers Unveiled as Fractal," *FOCUS*, **31** No. 2 (2011), 5–7.

[TATTERSALL] Tattersall, J., *Elementary Number Theory in Nine Chapters*, Cambridge University Press, Cambridge, England, 2005.

[YOUNG] Young, R. *Excursions in Calculus: An Interplay of the Continuous and the Discrete,* The Mathematical Association of America, Washington D.C., 1992.

16

Personalized Polynomials

PUZZLER: Dividing Consecutive Products

The product of any two consecutive integers is divisible by 2. This is obvious since one of them must be even.

Prove that the product of any three consecutive integers must be divisible by 6. (For example, $7 \times 8 \times 9 = 504$ is a multiple of 6.)

Prove that the product of any four consecutive integers is divisible by 24, the product of five consecutive integers by 120, and the product of six consecutive integers by 720. Does this remain true even if some of the consecutive integers are negative?

[By the way . . . What are the numbers 2, 6, 24, 120, and 720?]

Now let's consider sums:

The sum of any three consecutive numbers is divisible by 3. (For example, $15 + 16 + 17 = 48$, a multiple of three.) Can you see why?

Prove that the sum of 17 consecutive numbers is sure to be a multiple of 17. On the other hand, no sum of 18 consecutive integers is a multiple of 18. Prove this!

TIDBIT #1: Personalized Polynomials

My first name is JAMES and I am fond of the polynomial

$$p(x) = \frac{83}{24}x^4 - \frac{331}{12}x^3 + \frac{1657}{24}x^2 - \frac{647}{12}x + 10$$

because $p(0) = 10$ and the 10th letter of the alphabet is J, $p(1) = 1$ and the 1st letter of the alphabet is A, $p(2) = 13$ and the 13th letter of the alphabet is M, $p(3) = 5$ and the 5th letter of the alphabet is E, $p(4) = 19$ and the 19th letter of the alphabet is S.

Create a personalized polynomial that spells your name. (Read on to see how.)

When you succeed, prove that your polynomial always produces integer outputs for integer inputs, despite its fractional coefficients!

TIDBIT # 2: Foiling Intelligence Tests

Many intelligence tests ask questions of the ilk:

> What is the next number in the sequence?
>
> 2, 4, 6, 8, –

We can argue that the answer is 17 since the sequence is clearly(!) following the formula

$$p(n) = \frac{7}{24}n^4 - \frac{7}{4}n^3 + \frac{77}{24}n^2 + \frac{1}{4}n + 2.$$

[Check this. Show that $p(0) = 2$, $p(1) = 4$, $p(2) = 6$, $p(3) = 8$, and $p(4) = 17$.]

Or argue that it is -8 as the sequence could be defined by

$$q(n) = -\frac{3}{4}n^4 + \frac{9}{2}n^3 - \frac{33}{4}n^2 + \frac{13}{2}n + 2.$$

[Check this: Show that $q(0) = 2$, $q(1) = 4$, $q(2) = 6$, $q(3) = 8$ and $q(4) = -8$.]

The answer can be any value a you choose. We can claim that the sequence follows

$$f(n) = \frac{a - 10}{24} \cdot n^4 + \frac{10 - a}{4} \cdot n^3 + \frac{11a - 110}{24} \cdot n^2 + \frac{18 - a}{4} \cdot n + 2.$$

[Check this!]

Because we can create a polynomial to fit any data set shows that intelligence questions like these are not questions in mathematics.

How do we create a polynomial to give any set a desired next answer?

The French mathematician Joseph-Louis Lagrange (1736–1813) explored the same challenge and came up with what is today known as *Lagrange's interpolation formula*. We'll use it to find a polynomial that takes six specified values. Generalizations to a different number of values is straightforward.

Lagrange's Interpolation Formula. The polynomial that takes the values $a_0, a_1, a_2, a_3, a_4, a_5$ for $x = 0, 1, 2, 3, 4, 5$ in turn is:

$$
\begin{aligned}
p(x) = {} & a_0 \frac{(x - 1)(x - 2)(x - 3)(x - 4)(x - 5)}{(0 - 1)(0 - 2)(0 - 3)(0 - 4)(0 - 5)} \\
& + a_1 \frac{x(x - 2)(x - 3)(x - 4)(x - 5)}{1(1 - 2)(1 - 3)(1 - 4)(1 - 5)} \\
& + a_2 \frac{x(x - 1)(x - 3)(x - 4)(x - 5)}{2(2 - 1)(2 - 3)(2 - 4)(2 - 5)} \\
& + a_3 \frac{x(x - 1)(x - 2)(x - 4)(x - 5)}{3(3 - 1)(3 - 2)(3 - 4)(3 - 5)} \\
& + a_4 \frac{x(x - 1)(x - 2)(x - 3)(x - 5)}{4(4 - 1)(4 - 2)(4 - 3)(4 - 5)} \\
& + a_5 \frac{x(x - 1)(x - 2)(x - 3)(x - 4)}{5(5 - 1)(5 - 2)(5 - 3)(5 - 4)}.
\end{aligned}
$$

This looks complicated at first glance, but actually the polynomial is easy to understand.

There are six terms, one for each appearance of the six individual values $a_0, a_1, a_2, a_3, a_4, a_5$.

Each term has a numerator designed to vanish at all but one of the values $x = 0, 1, 2, 3, 4, 5$. The denominator of each term cancels with the numerator that arises when the numerator does not vanish so that the fraction has value 1. Multiplying by a_i causes the polynomial to have value a_i in that case. (To understand this, put $x = 2$ into the formula and see how that value a_2 appears.)

Exercise. Use Lagrange's interpolation formula to write the formula for a quadratic p with $p(0) = 40$, $p(1) = 3$, and $p(2) = 10$. Show that it gives $p(x) = 22x^2 - 59x + 40$.

Comment. Many algebra curricula have students find quadratics that fit three specified data points by solving a system of three linear equations in three unknowns. Lagrange's interpolation formula gives the answer more quickly.

Challenge. Prove that Lagrange's formula gives integer outputs for integer inputs. (This uses the opening puzzler: The product of any n consecutive integers—as appears in the numerator of each term of Lagrange's polynomial—is divisible by $n!$.)

Research Corner. Suppose $p(x)$ is a polynomial with integer coefficients. If $p(a) = 4$, $p(b) = 5$, and $p(c) = 4$ for some integers a, b and c, prove that a, b and c must be consecutive!

Suppose, instead, $p(x)$ takes the value 17 at four different integer inputs and the value 19 at another integer input. Prove that those five integers must be consecutive.

Establish other strange results about integer polynomials.

COMMENTARY, SOLUTIONS and THOUGHTS

Much of the content of this newsletter appears in [TANTON].

For any k consecutive positive integers $n + 1, n + 2, \ldots, n + k$ we have:

$$\frac{(n + 1)(n + 2) \cdots (n + k)}{k!} = \frac{(n + k)!}{n!k!} = \binom{n + k}{k}$$

which is a combinatorial coefficient (the number of ways to select k objects from $n + k$) and hence a whole number.

Challenge. What if some or all of the consecutive integers are negative?

Challenge. Prove, for positive integers n and k, that $(nk)!$ is sure to be a multiple of $(k!)^{n+1}$.

For sums we have:

$$(n + 1) + (n + 2) + \cdots + (n + k) = kn + \frac{k(k + 1)}{2} = \frac{1}{2}k(2n + k + 1).$$

If k is odd, then $\frac{2n+k+1}{2}$ is an integer and the sum of k consecutive integers is a multiple of k. This need not be the case if k is even. For example:

$$(n + 1) + (n + 2) + \cdots + (n + 18) = 9(2n + 19)$$

cannot be a multiple of 18.

Challenge. Actually, prove that "need not" should actually be "cannot." That is, show that $(n + 1) + (n + 2) + \cdots + (n + k)$ can never be a multiple of k if k is even.

Challenge. Twenty students sit in a circle. At the blow of the whistle all but one student moves clockwise to take a seat at a new position. Prove that two students move the same number of places.

Lagrange's Interpolation Formula

Given a sequence of n integers $a_0, a_1, \ldots, a_{n-1}$ Lagrange's method (see [ARCHER and WEISSTEIN] and the references therein) produces a degree $n - 1$ polynomial $p(x)$ that passes through each of the points (i, a_i). For an integer input, we see that a term in the construction of the polynomial is of the form

$$\frac{\text{a product of } a \text{ consecutive integers} \times \text{a product of } b \text{ consecutive integers}}{a! \times b!}$$

and so, by the opening remarks of this essay, is an integer. Thus Lagrange's polynomial produces integer outputs for integer inputs.

Comment. For those familiar with difference methods, there is a simpler way to construct Lagrange's interpolating polynomials. For example, to construct a polynomial that takes the values 10, 1, 13, 5, 19 ("James") at the values 0, 1, 2, 3, and 4, construct a difference table and examine its leading diagonal:

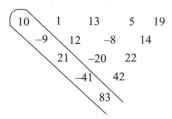

The desired polynomial is

$$p\left(n\right) = 10\binom{n}{0} - 9\binom{n}{1} + 21\binom{n}{2} - 41\binom{n}{3} + 83\binom{n}{4}$$

$$= 10 - 9n + \frac{21}{2}n(n - 1) - \frac{41}{6}n(n - 1)(n - 2) + \frac{83}{24}n(n - 1)(n - 2)(n - 3).$$

It is possible to use this method to construct a polynomial that passes through (most) any set of given points, not just those with the specific x-coordinates 0, 1, 2,.... For example, a polynomial that passes through the points $(3, 99)$, $(-5, 88)$, $(\pi, 1)$ is given by

$$p(x) = 99\frac{(x + 5)(x - \pi)}{8(3 - \pi)} + 88\frac{(x - 3)(x - \pi)}{2(2 - \pi)} + \frac{(x - 3)(x + 5)}{(\pi - 3)(\pi + 5)}.$$

Challenge.

a) There is no polynomial that passes through the points $(3, 8)$, $(6, 2)$ and $(3, 10)$. How does the interpolation formula break down?

b) The points $(5, 7)$, $(9, 10)$ and $(17, 16)$ are collinear. Does the interpolation formula detect this?

Final Thoughts on Integer Polynomials

To address some of the puzzlers presented in the "Research Corner" note that $x^n - a^n$ has factor $x - a$:

$$x^n - a^n = (x - a)(x^{n-1} + ax^{n-2} + a^2x^{n-2} + \cdots + a^{n-1}).$$

Thus, for any polynomial $p(x)$ we have that $x - a$ is a factor of $p(x) - p(a)$. In particular, if the polynomial has integer coefficients, then $p(a)$ and $p(b)$ are integers if a and b are, and $p(b) - p(a)$ is divisible by $b - a$ if a and b are distinct.

We'll put this observation to use!

1. Suppose that $p(x)$ has integer coefficients and $p(a) = 4$, $p(b) = 5$, and $p(c) = 4$ for distinct integer values a, b and c. Then $b - a$ is a factor of $5 - 4 = 1$ as is $c - a$. Thus $a = b \pm 1$ and $c = b \pm 1$, so a, b and c must be consecutive.

2. Suppose that $p(x)$ has integer coefficients and has value 17 at four different integer inputs a, b, c and d and the value 19 at another integer input e. Then $e - a$, $e - b$, $e - c$ and $e - d$ are distinct factors of 2, and so the integers a, b, c, d and e are consecutive (with e at the center!)

References

[ARCHER and WEISSTEIN] Archer, B., and Weisstein, E. W., "Lagrange interpolating polynomial." From *MathWorld*-A Wolfram Web Resource. URL: http://mathworld.wolfram.com/LagrangeInterpolatingPolynomial.html

[TANTON] Tanton, J. *THINKING MATHEMATICS! Volume 1: Arithmetic = Gateway to All*, www.lulu.com, 2009.

17

Playing with Pi

PUZZLER: A Rope Around the Earth

This puzzler is a classic:

> A rope fits snugly around the equator of the Earth. Ten feet is added to its length. When the extended rope is wrapped about the equator, it magically hovers at uniform height above the ground. How high off the ground?

> A second rope fits snugly about the equator of Mars and ten feet is added to its length. How high off the ground does this extended rope hover when wrapped about the planet's equator?

> A third rope fits snugly about the (tiny) equator of a planet the size of a pea. When ten feet is added to its length, how high off the equator does it hover?

Comment. The answers to these questions are surprising in three ways: they are the same, they can be computed with no knowledge of the radius of the planet under consideration, and the shared answer is surprisingly large, about 1.6 feet. (Adding just ten feet to a rope the length of the circumference of the Earth produces enough space under which to roll!)

This problem provides a wonderful activity for students. Using a length of rope as a radius, draw a circle on the ground with sidewalk chalk. Lay a long rope about its circumference and add ten feet to its length. Have a group of students—evenly spaced about the circle—attempt to wrap this extended rope about the original circle with a gap of constant width. When done, verify that the gap is about 19 inches. Repeat this activity with a very small circle drawn on the ground. It is a surprise to see the same gap of 19 inches appear!

There is nothing special about circles in these problems.

Exercise. A rope ten feet longer than the perimeter of a square is used to produce a concentric square. Verify that the gap between the two squares is 1.25 feet.

Exercise. A rope ten feet longer than the perimeter of an equilateral triangle is used to produce a concentric equilateral triangle. Verify that the gap between the triangles (to two decimal places) is 0.96 feet.

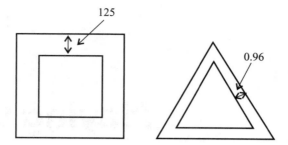

Thinking about Pi. Recall that π is defined to be the ratio of the circumference C of a circle to its diameter. If r is the radius, then

$$\pi_{\text{circle}} = \frac{C}{2r}.$$

Rearranging gives the formula $C = 2\pi r$. By subdividing the circle into "pizza wedges" and rearranging them to form a crude rectangle we obtain, in the limit of finer and finer wedges, the formula $A = \pi r^2$ for the area A of the circle.

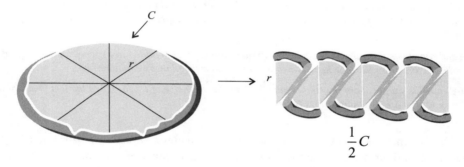

We can apply the same analysis to other shapes as well. Mimicking the definition "circumference over twice radius", define pi for a square, say, to be "perimeter over twice the radius of the square." There is a problem: a square offers different radii from which to choose.

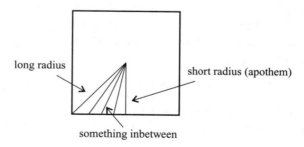

Since the apothem (short radius) is the easiest length to compute—it's just half the side length of the square—let's work with the apothem for our definition of pi. As we shall see this is the "correct" choice.

$$\pi_{\text{square}} = \frac{\text{perimeter of square}}{2 \times \text{apothem}}.$$

Exercise.

a)   Show that pi for a square is 4 for all squares.

b)   Show that the formula $C = 2\pi\, r$ works for a square. (Here π is pi for a square, C is the perimeter of the square, and r is the apothem.)

c)   Show that the formula $A = \pi\, r^2$ works for a square.

Continuing, let's define pi for an equilateral triangle to be

$$\pi_{\text{triangle}} = \frac{\text{perimeter of triangle}}{2 \times \text{apothem}}.$$

Exercise. Show that $\pi_{\text{triangle}} = 3\sqrt{3}$ for all equilateral triangles. Show that the formulas $C = 2\pi\, r$ and $A = \pi\, r^2$ hold for an equilateral triangle.

Define pi for a regular n-gon to be

$$\pi_{n\text{-gon}} = \frac{\text{perimeter}}{2 \times \text{apothem}}.$$

Exercise. Show that pi for a regular n-gon is $n \tan\left(\frac{180}{n}\right)$ and that the formulas $C = 2\pi r$ and $A = \pi r^2$ are still valid!

Exercise. A rope ten feet longer than the perimeter of a regular n-gon is wrapped about the polygon to produce a concentric regular n-gon. Verify that the gap between the polygons is $\frac{5}{\pi_{n\text{-gon}}}$ feet.

Research Corner. We can define a meaningful value of π for any polygon—not necessarily regular—that circumscribes a circle. Let r be the radius of the circle and P be the perimeter of the polygon. Set $\pi = \frac{P}{2r}$.

a)   A circle is inscribed in a right triangle with sides of length a, b and c, hypotenuse c.

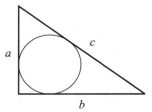

Show that the radius of the circle is $r = \frac{1}{2}(a + b - c)$. Show that if a, b, and c are integers, then r is too. Find the value of pi for this right triangle. Show that the formula $A = \pi r^2$ holds.

b)   Find r for the following trapezoid. What is its value of pi?

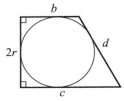

c)   Explore pi values for other types of polygons.

COMMENTARY, SOLUTIONS and THOUGHTS

It is a surprise that the gap between two concentric circles with the circumference of one ten feet longer than the circumference of the other is independent of the size of the circles.

If the circles have radii R_1 and R_2 with $R_1 < R_2$, then their circumferences are $C_1 = 2\pi R_1$ and $C_2 = 2\pi R_2$. From $C_2 - C_1 = 10$ we obtain $R_2 - R_1 = \frac{10}{2\pi} = \frac{5}{\pi} \approx 1.6$, independent of the radii.

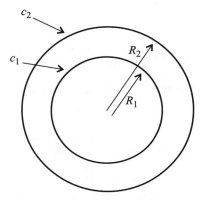

For a polygon with sides tangent to a circle of radius r, define π_{polygon} to be

$$\pi_{\text{polygon}} = \frac{P}{2r}$$

where P is the perimeter of the polygon.

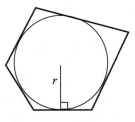

The definition is independent of the scale of the figure. (If the polygon is scaled by a factor k, say, then $P \to kP$ and $r \to kr$ and the ratio of "perimeter to twice inradius" is unchanged.)

For a square of side-length x we have $r = \frac{x}{2}$ and so $\pi_{\text{square}} = \frac{4x}{2\frac{x}{2}} = 4$.

For an equilateral triangle with incircle of radius r, its side-length $2r\sqrt{3}$ and $\pi_{\text{equilateral triangle}} = \frac{P}{2r} = \frac{6r\sqrt{3}}{2r} = 3\sqrt{3}$.

For a regular n-gon with incircle of radius r trigonometry shows that its side-length is $2r\tan(\frac{180°}{n})$ and so $\pi_{n\text{-gon}} = \frac{P}{2r} = \frac{n\cdot 2r\tan(180/n)}{2r} = n\tan\left(\frac{180°}{n}\right)$.

We have

Result 1. *For a polygon with perimeter P and area A, with sides tangent to a circle of radius r, $P = 2\pi_{\text{polygon}}r$ and $A = \pi_{\text{polygon}}r^2$.*

Proof. The first formula is just the definition of π_{polygon}. If the side-lengths of the polygon are $a_1, a_2, \ldots, a_n$ then its area is

$$A = \frac{1}{2}a_1r + \frac{1}{2}a_2r + \cdots + \frac{1}{2}a_nr = \frac{1}{2}Pr$$

from which the second formula follows.

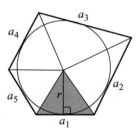

□

Result 2. *Consider a polygon with sides tangent to a circle. Suppose a rope ten units longer than the perimeter of the polygon is wrapped about the polygon to produce a concentric scaled version of the figure. Then the gap between the two figures is*

$$\frac{5}{\pi_{\text{polygon}}}.$$

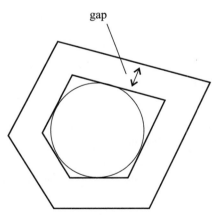

Proof. If the incircles of the two polygons have radii r_1 and r_2, with $r_1 < r_2$, then their perimeters are given by $P_1 = 2\pi_{\text{polygon}}r_1$ and $P_2 = 2\pi_{\text{polygon}}r_2$. From $P_2 - P_1 = 10$ we obtain $r_2 - r_1 = \frac{10}{2\pi_{\text{polygon}}} = \frac{5}{\pi_{\text{polygon}}}$.

□

So the notion of "pi" for a polygon is just as robust and as useful as that of "pi" for a circle!

Here is a third instance where the notion of pi for a polygon can be considered relevant.

Classic Wire Cutting Problems

The following problem appears in many calculus texts. It can be solved easily without calculus!

A wire of length 1 meter is to be cut into two pieces. The left piece is bent into a circle and the right piece into a square. Where along the wire should one make the cut to obtain two shapes with the smallest sum of areas?

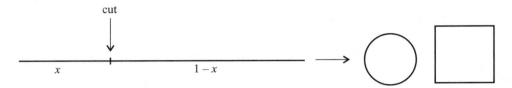

To solve the problem, call the length of the left piece of wire x. The length of the right piece is thus $1 - x$. Let a be the apothem (radius) of the circle, and b the apothem of the square. Let π_a be the value of pi for a circle and π_b its value of a square. Then $x = 2\pi_a a$ and $1 - x = 2\pi_b b$. We are being asked to minimize

$$\pi_a a^2 + \pi_b b^2.$$

In terms of x, the quantity we seek to minimize is

$$\pi_a \left(\frac{x}{2\pi_a}\right)^2 + \pi_b \left(\frac{1-x}{2\pi_b}\right)^2 = \frac{\left(\frac{1}{\pi_a} + \frac{1}{\pi_b}\right)x^2 - \frac{2}{\pi_b}x + \frac{1}{\pi_b}}{4}.$$

This is a parabola, concave up, with vertex (and hence minimum) occurring at

$$x = -\frac{\left(-\frac{2}{\pi_b}\right)}{2\left(\frac{1}{\pi_a} + \frac{1}{\pi_b}\right)} = \frac{\pi_a}{\pi_a + \pi_b}.$$

For this x we have $1 - x = \frac{\pi_b}{\pi_a + \pi_b}$.

Thus one should cut the wire at the position such that the ratio of the two lengths, $\frac{x}{1-x}$, equals the ratio of the values of pi for the two shapes: $\frac{\pi_a}{\pi_b}$.

This generalizes:

Given any two polygons with incircles, cutting the wire at a position yielding pieces of lengths in the ratio of the pi-values for those shapes minimizes total area.

Challenge. Show that the ratios of the areas of the two shapes at this optimal position is also $\frac{\pi_a}{\pi_b}$.

Challenge. We can solve this problem with calculus. We have the equation $2\pi_a a + 2\pi_b b = 1$ and we wish to minimize $A = \pi_a a^2 + \pi_b b^2$. Differentiate the first equation with respect to a to obtain $\frac{db}{da} = -\frac{\pi_a}{\pi_b}$. At its minimum, $\frac{dA}{da} = 0$. Use this to show that $a = b$. Thus the two shapes have the same apothem at the optimal position and hence the ratio of their circumferences (and areas, for that matter) is $\frac{\pi_a}{\pi_b}$.

Challenge. A wire, one meter long, is to be cut into N pieces and each piece is to be formed into a specified shape—either a circle or some polygon with a well-defined incircle. Show that cutting the wire into sections of lengths in the same proportions of the pi-values for the shapes minimizes the sum of areas of the shapes.

This phenomenon was also observed by P. Cade and R. A. Gordon in [CADE and GORDON]. They phrase their work in terms of apothem values rather than pi-values.

Final Tidbit

The inradius r of a right triangle with integer side-lengths a, b, and c is itself an integer. (Assume c is the hypotenuse.)

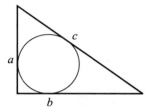

To see this, the area of the triangle can be computed as $\frac{1}{2}ab$ and as $\frac{1}{2}(a + b + c)r$. Thus

$$r = \frac{ab}{a + b + c}.$$

Multiplying the numerator and denominator by $a + b - c$, and using $a^2 + b^2 = c^2$, yields

$$r = \frac{ab(a + b - c)}{a^2 + b^2 + 2ab - c^2} = \frac{1}{2}(a + b - c).$$

If $a^2 + b^2 = c^2$ it is impossible that just one of the numbers a, b, c is odd or that all three are odd. (For instance, a odd, b even, forces c^2, and hence c, to be odd.) Thus $a + b - c$ is even, making r an integer.

Furthermore, $\pi_{\text{right triangle}} = \dfrac{P}{2r} = \dfrac{a + b + c}{a + b - c}.$

Challenge. Show that the value of pi for this trapezoid is $\pi_{\text{trapezoid}} = \dfrac{2b + 2c}{b + c - d}.$

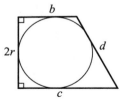

See [TANTON] for more curious ways to work with pi.

Reference

[CADE and GORDON] Cade, P., and Gordon, R. A., An apothem apparently appears, *College Mathematics Journal*, **36** (2005), 52–55.

[TANTON] Tanton, J. *Weird Ways to Work with Pi*, www.lulu.com, 2012.

18

Pythagoras's Theorem

PUZZLER: Pythagorean Surprise

A triple of positive integers (a, b, c) is called a *Pythagorean triple* if $a^2 + b^2 = c^2$.

Did you know that an ordinary multiplication table contains such triples?

×	1	2	3	4	5	6	7
1	1	2	3	4	5	6	7
2	2	④	6	8	⑩	12	14
3	3	6	9	12	15	18	21
4	4	8	12	16	20	24	28
5	5	⑩	15	20	㉕	30	35
6	6	12	18	24	30	36	42
7	7	14	21	28	35	42	49

Choose two numbers on the diagonal (these are square numbers) and two numbers to make a square. Sum the two square numbers, take their difference and sum the other two numbers. You now have a Pythagorean triple!

$$a = 25 - 4 = 21$$
$$b = 10 + 10 = 20 \quad \Rightarrow \quad 20^2 + 21^2 = 29^2$$
$$c = 25 + 4 = 29$$

Choosing 36 and 1 gives

$$a = 36 - 1 = 35$$
$$b = 6 + 6 = 12 \quad \Rightarrow \quad 12^2 + 35^2 = 37^2$$
$$c = 36 + 1 = 37$$

Question 1. Which two square numbers give the triple $(3, 4, 5)$? Which give $(5, 12, 13)$ and $(7, 24, 25)$?

137

Question 2. Why does this work?

Tough Challenge. This method fails to yield $(9, 12, 15)$. But if we divide by 3 we have the triple $(3, 4, 5)$, which is obtainable. Show that every Pythagorean triple with no common factor among its terms (other than 1) appears.

TIDBIT: Pythagoras Meets the Third Dimension

A classic problem in geometry is to work out the length of the longest diagonal in a rectangular box. For example, what is the distance between points A and B in this box?

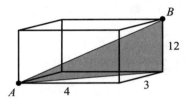

The Pythagorean theorem shows that the length of the diagonal on the base is 5 units long. Using it on the shaded triangle yields the length we seek: 13 units.

In general, this shows that the length of the longest diagonal d in an $a \times b \times c$ rectangular box satisfies:

$$d^2 = a^2 + b^2 + c^2.$$

This is one way to think of the Pythagorean theorem extended to the third dimension, but it is not the only way!

Consider the triangle inside a rectangular box as shown. What is its area?

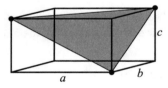

We can compute this with some ease using some algebra.

To give a clearer view remove the three right triangles about the top front corner of the box. Draw an altitude for the triangle and label the lengths x, y, and h as shown.

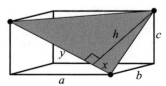

Do you see that $x^2 + h^2 = b^2 + c^2$ and $y + x = \sqrt{a^2 + c^2}$? Do you also see that $y^2 + h^2 = a^2 + b^2$?

Let's use the third equation with the first two. We have:

$$\left(\sqrt{a^2 + c^2} - x\right)^2 + h^2 = a^2 + b^2$$

so

$$a^2 + c^2 - 2x\sqrt{a^2 + c^2} + x^2 + h^2 = a^2 + b^2$$

yielding

$$2x\sqrt{a^2 + b^2} = x^2 + h^2 + c^2 - b^2 = 2c^2.$$

Thus

$$x = \frac{c^2}{\sqrt{a^2 + c^2}}.$$

Solve for h^2

$$h^2 = b^2 + c^2 - x^2 = b^2 + c^2 - \frac{c^4}{a^2 + c^2}$$
$$= \frac{a^2 b^2 + b^2 c^2 + a^2 c^2}{a^2 + c^2}.$$

(Heavens!)

We are now set to compute the area of the triangle: $A = \frac{1}{2} \times \sqrt{a^2 + c^2} \times h$. ("Half base times height.") To avoid square roots let's instead compute A^2:

$$A^2 = \frac{1}{4}(a^2 + c^2)h^2 = \frac{1}{4}(a^2 b^2 + b^2 c^2 + a^2 c^2)$$
$$= \left(\frac{1}{2}ab\right)^2 + \left(\frac{1}{2}bc\right)^2 + \left(\frac{1}{2}ac\right)^2.$$

Each term squared is the area of one of the right triangles on the side of the rectangular box, those that we removed from the top front corner of the box. We have:

A triangle of area D sits across three mutually perpendicular right triangles. If these triangles have areas A, B, and C, then $A^2 + B^2 + C^2 = D^2$.

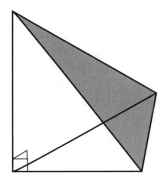

Exercise. A triangle has vertices on the x-, y- and z-axes at positions 3, 5 and 8 respectively. What is its area?

This alternative 3-D version of Pythagoras's theorem is not particularly well known.

Research Corner. The pictured box has the property that each side-length, and each diagonal across a face, has integer length.

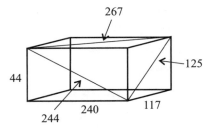

Challenge 1. Find another example.

Challenge 2. The length of the longest diagonal inside this box, $\sqrt{73225}$, is not an integer. No one knows an example of a box with all side lengths and *all* diagonals integers. For world fame, can you find one?

COMMENTARY, SOLUTIONS and THOUGHTS

If we pick square numbers m^2 and n^2 from rows m and n from a multiplication table, $m > n$, it is not difficult to see that setting

$$a = m^2 - n^2,$$
$$b = 2mn,$$
$$c = m^2 + n^2$$

yields a Pythagorean triple:

$$a^2 + b^2 = (m^2 - n^2)^2 + (2mn)^2$$
$$= m^4 + n^4 - 2m^2n^2 + 4m^2n^2$$
$$= m^4 + n^4 + 2m^2n^2$$
$$= (m^2 + n^2)^2$$
$$= c^2.$$

The difficulty lies in proving that every Pythagorean triple arises in this way. The following result is given in book X of Euclid's *Elements* ([TATTERSALL1]) written ca. 300 B.C.E.:

Theorem: (Classification of Primitive Pythagorean Triples). *If (a, b, c) is a primitive Pythagorean triple (that is, a triple of integers with greatest common factor 1 satisfying $a^2 + b^2 = c^2$), then there are integers m and n so that $a = m^2 - n^2$, $b = 2mn$ and $c = m^2 + n^2$.*

The proof uses the following key property of prime numbers:

If a prime p is a factor of MN, then either p is a factor of M or a factor of N.

We prove the theorem in steps:

1. *At least one of a, b, and c is odd.* They cannot all be even as they do not share a common factor other than 1.

2. *At least one of a and b is odd.* If both are even, then $c^2 = a^2 + b^2$ is even, making c even, impossible in a primitive triple.

3. *Precisely one of a and b is odd.* If both are odd then $a = 2k + 1$ and $b = 2r + 1$ for some integers k and r, so $a^2 + b^2 = (2k + 1)^2 + (2r + 1)^2 = 4(k^2 + r^2 + k + r) + 2$. That is, c^2 is two more than a multiple of four. But this is impossible! Either c is even, so c^2 is a multiple of four, or c is odd, so c^2 is one more than a multiple of four. (If $c = 2s + 1$, then $c^2 = 4s^2 + 4s + 1$.)

Without loss of generality, we assume that a is odd and b is even. It follows that c^2, and hence c, is odd.

4. Write $b^2 = c^2 - a^2 = (c - a)(c + a)$. As b is even it is divisible by two. Thus

$$\left(\frac{b}{2}\right)^2 = \left(\frac{c - a}{2}\right)\left(\frac{c + a}{2}\right)$$

is an equation involving integers because a and c are odd.

To simplify the notation, set $B = \frac{b}{2}$, $N = \frac{c-a}{2}$ and $M = \frac{c+a}{2}$. We have $B^2 = N \cdot M$.

5. *The integers N and M share no prime factor*: If p is a common prime factor, then p divides into their sum $N + M = \frac{c-a}{2} + \frac{c+a}{2} = c$ and their difference $M - N = \frac{c+a}{2} - \frac{c-a}{2} = a$, and hence into $b^2 = c^2 - a^2$, contradicting the fact that the triple (a, b, c) possesses no common prime factor.

6. We have now two integers N and M sharing no prime factor with the property that their product is a square number, namely, B^2. Any prime that appears in the prime factorization of B appears twice in $B \cdot B = B^2$, and so appears twice either in N or in M. Thus all the primes that appear in the prime factorization of N and of M do so an even number of times. This means N and M are each square numbers

$$N = n^2, \quad M = m^2$$

for some integers n and m.

We're now done. Adding and subtracting $\frac{c-a}{2} = n^2$ and $\frac{c+a}{2} = m^2$ gives

$$c = m^2 + n^2,$$
$$a = m^2 - n^2.$$

Solving for b^2 gives $b^2 = c^2 - a^2 = 4m^2n^2$ and so

$$b = 2mn. \qquad \square$$

Comment. I sent the following puzzler out to all the Math Institute attendees.

3-4-5 Wannabies. *A **Pythagorean triple** is a set of three positive integers a, b, c satisfying the relation $a^2 + b^2 = c^2$. These are the side-lengths of a right triangle with integer sides.*

Many students memorize the 3-4-5 triple ($3^2 + 4^2$ and 5^2 are both 25), but there are also

$$5\text{-}12\text{-}13 \quad 7\text{-}24\text{-}25 \quad 20\text{-}21\text{-}29 \quad 119\text{-}120\text{-}169 \quad 51\text{-}140\text{-}149$$

$$39\text{-}80\text{-}89 \quad 240\text{-}782\text{-}818 \quad 2295\text{-}288\text{-}2313 \quad 13205\text{-}5148\text{-}14173$$

and many more!

In each example at least one term is divisible by 3, one term divisible by 4, and one term divisible by 5! (For example, in 39-80-89, 39 is divisible by three, 80 is divisible by four, and 80 is divisible by five.)

Is this always true? Must a Pythagorean triple possess terms divisible by 3, and by 4 and by 5? If so, why? If not . . . find an example of a triple that does not possess this property.

Care to take on the challenge?

Another approach to deriving a parameterization of Pythagorean triples is to use complex numbers:

If $c^2 = a^2 + b^2$, then $c^2 = (a + ib)(a - ib)$. Let $m + in$ be a square root of $a + ib$. Then $m - in$ is a square root of $a - ib$ and $c = (m + in)(m - in) = m^2 + n^2$. From $(m + in)^2 = a + ib$ we obtain $a = m^2 - n^2$ and $b = 2mn$.

The challenge here is to prove that m and n are integers (and can be taken to be positive integers).

Challenge. Do so! Prove that if (a, b, c) is a triple of integers sharing no common factor and satisfying $a^2 + b^2 = c^2$, then the complex square root $m + in$ of $a + ib$ has integer components.

Pythagorean Strings

The newsletter presented an example of a Pythagorean quadruple: four integers (a, b, c, d) satisfying $a^2 + b^2 + c^2 = d^2$. Other examples include the quadruples

$$(1, 2, 2, 3), \quad (2, 3, 6, 7), \quad (1, 4, 8, 9), \quad (6, 6, 7, 11).$$

High school students in my 2008 Advanced Topics course pointed out that there are infinitely many examples of such quadruples because

$$x^2 + (x + 1)^2 + \left(x^2 + x\right)^2 = \left(x^2 + x + 1\right)^2.$$

This does not generate all examples, however, and no one has found a formula that generates all Pythagorean quadruples ([WEISSTEIN2]).

There exist Pythagorean quintuplets—$(3, 4, 12, 84, 85)$, for example—and hextuplets, and so on. In fact

Theorem. *There exists a sequence of natural numbers a_1, a_2, a_3, . . . such that, for each n, $a_1^2 + a_2^2 + \cdots + a_n^2$ is a square.*

Proof. The square numbers 1, 4, 9, 16, ... differ by successive odd numbers, so for any odd number a different from 1, it is possible to find two consecutive squares m^2 and $(m + 1)^2$, such that

$$a = (m + 1)^2 - m^2.$$

Following the generating formula for Pythagorean triples, set

$$b = 2m(m + 1) \text{ and } c = (m + 1)^2 + m^2.$$

Because c is odd we have shown

Every odd number a (different from 1) is part of a triple $a^2 + b^2 = c^2$ with c odd.

And we have a method for finding the triple. We can apply the same algorithm to the odd number c to show that it is part of a triple (c, d, e) with e odd, and then to the odd number e, and so on.

For example, starting with

$$3^2 + 4^2 = 5^2$$

we obtain

$$5^2 + 12^2 = 13^2, \quad 13^2 + 84^2 = 85^2, \quad 85^2 + 3612^2 = 3613^2$$

and so on. We can combine them:

$$3^2 + 4^2 = 5^2$$
$$3^2 + 4^2 + 12^2 = 13^2$$
$$3^2 + 4^2 + 12^2 + 84^2 = 85^2$$
$$3^2 + 4^2 + 12^2 + 84^2 + 3612^2 = 3613^2.$$

Thus we generate an infinite string 3, 4, 12, 84, 3612, ... with the desired property. □

De Gua's Theorem

According to E. Weisstein ([WEISSTEIN1]), the extension of Pythagoras's result to trirectangular tetrahedra is due to J.P. de Gua de Malves (1712–1785). The distance formula of a point from a plane gives a swifter means for deriving the result.

Consider the plane $\frac{x}{a} + \frac{y}{b} + \frac{z}{c} = 1$, which crosses the axes at positions a, b and c, and the trirectangular tetrahedron that has the origin and $(a, 0, 0)$, $(0, b, 0)$ and $(0, 0, c)$ as its vertices. Its volume is $\frac{1}{3}\left(\frac{1}{2}ab\right)c = \frac{1}{6}abc$.

If h represents the distance of the origin from the plane $\frac{x}{a} + \frac{y}{b} + \frac{z}{c} = 1$ and D the area of the triangular face in this plane, then the volume of the solid is also given by $\frac{1}{3}Dh$.

The distance formula shows

$$h = \frac{\left|\frac{0}{a} + \frac{0}{b} + \frac{0}{c} - 1\right|}{\sqrt{\frac{1}{a^2} + \frac{1}{b^2} + \frac{1}{c^2}}} = \frac{1}{\sqrt{\frac{1}{a^2} + \frac{1}{b^2} + \frac{1}{c^2}}}$$

and so from $\frac{1}{6}abc = \frac{1}{3}Dh$ it follows that

$$4D^2 = a^2b^2 + a^2c^2 + b^2c^2,$$

equivalent to de Gua's result.

Euler's Brick

Swiss mathematician Leonhard Euler (1707–1783) investigated the existence of rectangular boxes with integer side-lengths, integer face diagonals, and an integer space diagonal. That is, he sought a triple of integers (a, b, c) with the property that each of

$$\sqrt{a^2 + b^2}, \quad \sqrt{a^2 + c^2}, \quad \sqrt{b^2 + c^2} \text{ and } \sqrt{a^2 + b^2 + c^2}$$

is an integer. He found no such example, nor has anyone to this day.

The example presented in the newsletter, $(240, 117, 44)$ was first discovered in 1719 by German mathematician Paul Halcke ([PETERSON]), and has all but the space diagonal integral. It is the smallest example with integer face diagonals.

There are infinitely many examples of bricks (a, b, c) with $\sqrt{a^2 + b^2}$, $\sqrt{a^2 + c^2}$, and $\sqrt{b^2 + c^2}$ integral. Let (p, q, r) be any Pythagorean triple and set

$$a = p\left(4q^2 - r^2\right), \quad b = q\left(4p^2 - r^2\right), \quad c = 4pqr.$$

Then

$$\sqrt{a^2 + b^2} = r^3, \quad \sqrt{a^2 + c^2} = p\left(r^2 + 4q^2\right), \text{ and } \sqrt{b^2 + c^2} = q\left(r^2 + 4p^2\right)$$

are all integers. This family is due to Nicolas Saunderson (1682–1739). ([TATTERSALL2]). It does not generate all examples of bricks with integral face diagonals, missing $(720, 132, 85)$, for example.

The triple $(672, 153, 104)$ has two face diagonals and the space diagonal integral, but the third face diagonal is not an integer.

The triple $\left(18720, 7800, \sqrt{211773121}\right)$ has all three face diagonals and the space diagonal integral (but, of course, one of its side lengths is not!)

The search for a "perfect cuboid"—or a proof that one does not exists—continues.

Challenge. Suppose (a, b, c) are the integer sides of a brick with integer face diagonals. Prove that at least one of the numbers a, b, c is divisible by 11.

References

[TANTON] Tanton, J., *Geometry Vol. 2*, www.jamestanton.com, 2009.

[TATTERSALL1] Tattersall, J., *Elementary Number Theory in Nine Chapters*, 2nd ed., Cambridge University Press, Cambridge, England, 2005.

[TATTERSALL2] Tattersall, J., Nicholas Saunderson: The blind Lucasion Professor, *Historia Mathematica*, **19** (4) (1992), 356–370.

[PETERSON] Peterson, I., Euler's bricks and perfect polyhedra in *Ivars Peterson's Math Trek* (October 1999). URL: www.maa.org/mathland/mathtrek_10_25_99.html

[WEISSTEIN1] Weisstein, E.W., De Gua's theorem. From *MathWorld*-A Wolfram Web Resource. URL: http://mathworld.wolfram.com/deGuasTheorem.html

[WEISSTEIN2] Weisstein, E.W., Pythagorean quadruple. From *MathWorld*-A Wolfram Web Resource. URL: http://mathworld.wolfram.com/PythagoreanQuadruple.html

19

On Reflection

PUZZLER: Pocketing the Billiard Ball

A popular discovery activity in the middle school curriculum is

A ball is shot from the bottom left corner of a 3 × 5 billiard table at a 45 degree angle. The ball traverses the diagonals of individual squares drawn on the table, bouncing off the sides of the table at equal angles. Into which pocket, A, B, or C, will it eventually fall?

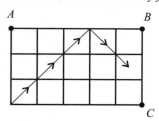

Experiment with the tables pictured below. What do you notice about those tables that have the ball fall into the top-left pocket A? Into the top-right pocket B? Into the bottom-right pocket C? Test your theories with more tables.

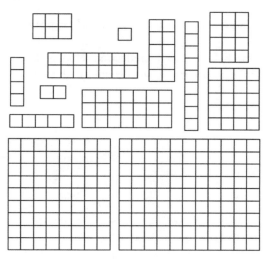

Experimentation shows that a ball bouncing in an $n \times m$ table with both n and m odd ends in pocket B, with n even and m odd in pocket A, and with n odd and m even in pocket C. Matters are more obscure if both n and m are even. (You may have noticed that I avoided such tables in the diagrams.)

Often the investigation ends here, but there are some interesting questions that could be asked—and answered!

Question 1. Must every ball land in pocket A, B, or C? Is it possible for a ball to return to where it started? Is it possible for a ball to enter an infinite loop and never fall into a pocket?

Question 2. Color the cells of each table black and red like a checkerboard. What do you notice about the path of the ball across the red cells? Across the black cells? Imagine (do not draw!) a 13×29 table. Use what you observe to explain why the ball can possibly fall only into pocket B. Explain the observations made at the top of this column.

Question 3. Did you notice that the ball passes through each and every cell for each of the tables above? Not every billiard table has this property. Find a condition on the numbers n and m that ensures that this will be the case for an $n \times m$ table.

Question 4. Develop a theory that will predict into which pocket the ball will fall if both n and m are even.

TIDBIT: The Same Problem Twice

Answering the same question in more than one way can be illuminating. Behold!

A dog, starting at point A wishes to walk to point B via a path that first visits a wall. (Assume the dog will walk along two straight line segments to do this.)

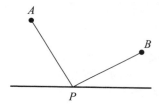

Locate the point P on the wall that gives the shortest path from A to P to B.

Answer 1. Here's a sneaky trick. Let B' be the reflection of point B on the other side of the wall. Then a path from A to P to B is matched by a path of equal length from A to P to B', and vice versa.

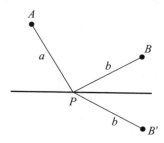

Clearly the shortest path from A to B' is the straight line path, making equal angles as shown below. Consequently the shortest path from A to B via the wall is the one that bounces off the wall at equal angles.

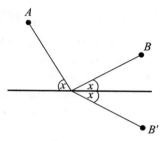

Answer 2. An ellipse with foci points A and B has the property that all points P on the curve give the same path length A to P to B: $AP + PB =$ a fixed value.

Draw an ellipse about the points A and B. If the ellipse is too small it won't reach the wall and give an appropriate dog walking path. If it is too big, it can be replaced by a smaller ellipse giving a shorter path from A to P to B. Thus the point P we seek is the location where the smallest ellipse possible just touches the wall. That is, the wall represents a tangent line to an ellipse and P is the point of contact.

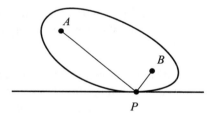

In both answers P must be the same. We have thus proven the famous reflection property of an ellipse! **Any path from one focus of an ellipse to the other via a point on the ellipse makes equal angles to the tangent line to the ellipse.** Thus a ball thrown from one focus to any point on the wall of the ellipse will head directly through the second focus. Sound waves in the elliptical Mormon Tabernacle in Salt Lake City and the Whispering Gallery in the United States Capitol building in Washington D.C. operate this way! (The sound waves will bounce back and forth infinitely often between the two foci!)

Research Corner: The Other Reflection Properties? Devise other problems that can be solved in two different ways to prove the famous reflection properties of the parabola and of the hyperbola.

COMMENTARY, SOLUTIONS and THOUGHTS

This billiard activity is simply super! (See [NCTM] and [LAPPAN].) It provides many avenues of exploration for students at all levels. Exploring billiard action on general polygonal-shaped tables is part of current active research! (See [De MARCO], for instance.)

The questions I pose in the newsletter are not usually examined at the school level. When given permission to think openly students of all ages will ask them.

Question 1. *Must every ball land in either pocket A, B, or C. Is it possible for a ball to return to where it started? Is it possible for a ball to enter an infinite loop and never fall into a pocket?*

The motion of the ball is time reversible: If the ball is currently traversing one particular cell (in a particular direction), then there is no doubt from which cell it just came. This observation is key.

Could the ball enter the same square twice in the same direction? No! Each time the ball enters a repeat square it must have come from a previous repeat square. There is no first cell the ball can visit twice.

Could the ball enter the same square twice in the opposite directions? No! To do this the ball must have traversed a previous square twice in opposite directions, and so, again, there is no first cell the ball can traverse twice in this way. In particular, the ball will never return to the bottom left cell. (Could a ball traverse the same square but along different diagonals?)

As there are only finitely many cells in the grid and no cell can be visited twice along the same diagonal, the ball's path must terminate. This establishes that the ball is sure to fall into one of the pockets.

Question 2. *Color the cells of each table black and red like a checkerboard. What do you notice about the path of the ball across the red cells? Across the black cells? Imagine (do not draw!) a 13 × 29 table. Use what you observe to explain why the ball can only possibly fall into pocket B. Explain the observations made in the paragraph at the top of this column.*

This question suggests a popular approach to analyzing billiard motion because the ball traverses red cells only along northwest diagonals and the black cells along southwest diagonals (or vice versa), and matters about the ball's behavior fall into place.

But analysis is easier if we color the grid points of the table two colors in a checkerboard pattern. (We've chosen the colors black and white here.)

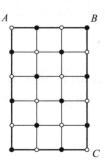

If we assume that the start corner is colored black, then it is clear that the ball will only visit black grid points. We also see that

If n and m are both odd, then pockets A and C will be white and pocket B black. The ball must fall in pocket B.

If n is even and m odd, then only pocket A is black. The ball must fall there.

If n is odd and m even, only pocket C is black and the ball must fall there.

When n and m are both even, each corner cell is black. This case requires further analysis, which we shall conduct in a moment.

Question 3. *Did you notice that the ball passed through each and every cell in each of the tables above? Not every billiard table has this property. Find a condition on the numbers n and m that ensures that this will be the case for an n × m table.*

Let N be the number of cells visited by the bouncing ball. We certainly have $N \leq nm$.

The ball moves left and right (as well as up and down) about the table to fall eventually into one of the pockets on the left or the right of the table. Looking at only the horizontal component of the motion of the ball, it is clear that N must be a multiple of m. Similarly, looking at the vertical component of the ball's motion, N must also be a multiple of n. If n and m are coprime (that is, share no common factors), then we have that N is a multiple of mn. Since $N \leq mn$ we conclude that $N = mn$.

Comment. As we shall see next, if the numbers m and n are not coprime (if both are even, for instance), then some cells will be missed.

Question 4. *Develop a theory that will predict into which pocket the ball will fall if both m and n are even.*

A 4×6 table, for example, is a 2×3 table in disguise and so the ball will fall into pocket A.

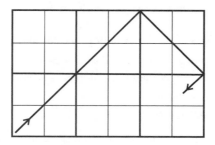

A 10×50 table is a 5×25 table in disguise (pocket B), and a 200×4000 table is a 1×20 table in disguise (pocket C).

In general, if n and m have greatest common factor d,

$$n = ad \quad \text{and} \quad m = bd$$

and an $n \times m$ table is an $a \times b$ table in disguise with a and b coprime. The ball will traverse precisely abd cells (why this number?) and will land into pocket A, B, or C according to whether a and b are odd or even as dictated before.

More Billiards

The mathematics behind the billiards game is surprisingly rich.

Suppose we start the billiard ball at *any* grid point of the table, setting it in motion along the diagonal of a cell. Assume now that all four corners represent pockets.

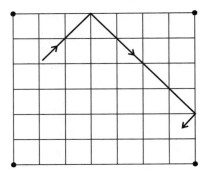

In playing on a 6×7 table, for example, we get the impression that every starting point and starting direction yields a path ending in a pocket. Is the same true for a 14×25 table? A 6×8 table? What can one say about the behavior of billiard balls on general $n \times m$ grids? Perhaps look at some examples before reading on.

There are many subtleties in this exploration and a complete analysis requires some serious effort. We present here partial results. Assume we have colored the grid points of an $n \times m$ table in a black and white checkerboard pattern, with the bottom-left corner black.

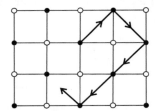

1. The ball only visits grid-points of the same color as its starting grid-point.

2. If a ball starts in a corner, it will terminate in a corner. Moreover, if $\gcd(n, m) = 1$, the ball will fall into the only remaining pocket of the same color as its starting pocket and will pass through every cell of the table before doing so.

(We have essentially already established this.)

3. If n and m are both even, then a ball that starts at a white grid-point enters an infinite loop.

4. If $\gcd(n, m) = 2$ and the ball starts at a white grid-point, then the ball enters an infinite loop that traverses every cell of the table.

5. If $\gcd(n, m) = 2$, then any ball starting at a black grid-point will fall into a corner.

Challenge. Prove these claims.

On 4×4 and 4×8 tables there are infinite loops that start at a white grid-point and do not pass through every cell of the table, and paths that start at black grid-points that do not end at corners. The condition that $\gcd(n, m) = 2$ is important.

As this is an incomplete list of results, what then is the ultimate theorem that completely describes diagonal billiard-ball motion in rectangular tables?

Reflection Properties of Conics

We can go far in solving optimization problems using reflections. For example, consider the following dual problem to the dog-walking problem:

Two points A and B lie on opposite sides of a line with point A closer to the wall than point B.

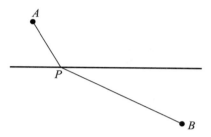

Find the location of the point P on the line that maximizes the difference $PB - PA$.

Answer 1. Reflect A across the line to A'. Let P be the point on the line such that P, A', and B are collinear. Then $PB - PA$ is the length x.

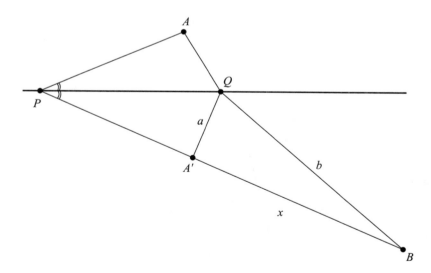

For any other point Q on the line, the analogous quantity is the difference of lengths $b - a$. By the triangle inequality, $x + a > b$, and so $x > b - a$. Thus P gives the maximal value and is the point we seek.

This solution, as in the first problem, relied on constructing the straight line between one of the points A and B and the reflection of the other. It too produces congruent angles.

Answer 2. A hyperbola with foci A and B is the locus of points P with the property that $PB - PA$ has a fixed value. The largest value it can be is AB and the smallest it can be is $-AB$. The cases with $PB - PA$ positive or negative give the two branches of the hyperbola and the case $PB - PA = 0$ corresponds to the degenerate example of a single straight line, the perpendicular bisector of $\overline{AB}$.

Consider the locus of points P satisfying $PB - PA = k$ for k with $0 \le k \le AB$. This set of points is the branch of a hyperbola that wraps around A, with $k = 0$ yielding the straight perpendicular bisector. As k increases, we can find the last branch to touch the line. It gives the largest possible value of $PB - PA$ for points on the line and is tangent to the line.

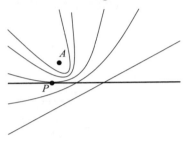

But we know that P is the point obtained by extending the line segment $\overline{A'B}$ and that congruent angles appear. We have established the reflection property of the hyperbola:

A ray of light directed from one focus of a hyperbola reflects off the hyperbola directly away from the second focus.

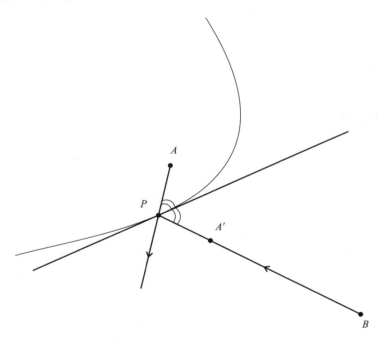

We can go further. Let A and B be two points in the plane. Then there is an infinite family of ellipses with A and B as foci and an infinite family of hyperbolas, also with A and B as foci.

Let P be an arbitrary point and consider the tangent line to an ellipse with foci A and B that passes through P. P solves the minimization problem (the dog-walking problem) for that line and we have congruent angles.

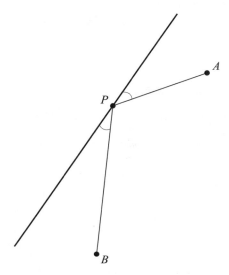

If we draw a perpendicular line through P we see that the same diagram solves the maximization problem that defines the reflection property of a hyperbola!

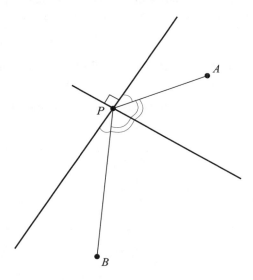

This establishes that our two infinite families of ellipses and hyperbolae sharing the same foci are orthogonal: whenever one of each curve intersect at a point P their tangent lines there are perpendicular.

Having now analyzed the ellipse and the hyperbola, we ask

Is there a minimization or maximization problem that identifies the reflection property of a parabola?

(For the study of ellipses we considered two points *A* and *B* on the same side of a line, and for hyperbolas points *A* and *B* on opposite sides of a line. What is the "intermediate" situation?)

Comment. The reflection properties of conics can also be explored by activities in folding paper. See newsletter 7 for more about this, or Chapter 29 of [TANTON2].

Other Reflection Challenges

The dog-walking challenge can be posed another way:

> *Abigail and Beatrice stand at A and B, respectively, in a squash court. Abigail wishes to hit a ball against the wall so that it bounces off and heads directly towards Beatrice. Describe the location of P on the wall towards which Abigail should aim to accomplish this. (We assume that air resistance, friction, and inelasticity have no effect on the motion of bouncing balls.)*

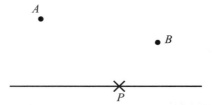

The point *P* that offers a path with angle of incidence matching angle of reflection does the trick. As we have seen, Abigail can find the point by imagining that the wall is a mirror and aiming for Beatrice's reflection.

Challenge. Explain why, in ideal circumstances, a ball or light ray bouncing from a (mirrored) wall does so with equal angles of reflection.

Young students typically delight in thinking about a more complicated version of the previous challenge:

> *Abigail and Beatrice again stand A and B, respectively, in a squash court, but this time Abigail wishes to make use of all three walls of the court. She wants to hit a ball against wall 1 so that it then bounces off wall 2 to hit wall 3 and then head directly toward Beatrice. Describe the location of a point P on wall 1 towards which Abigail should aim to accomplish this.*

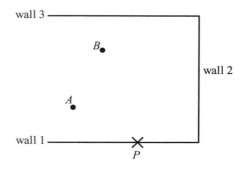

Abigail solves this by aiming for the reflection of the reflection of the reflection of Beatrice!
    This problem has some rich variations.

Challenge. Given a point in the interior of a rectangular billiards table describe how to hit the ball so that it returns to where it started.

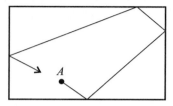

Under what conditions, if the ball were allowed to stay in motion, does it retrace the same quadrilateral path over and over again in an infinite loop?

Challenge. Suppose we are given an acute triangle.

a)   Show that there is an inscribed triangle with one vertex on each side of the original triangle of shortest perimeter.

[This problem was first posed, and solved, by an Italian priest and scholar Giovanni Fagnano (1715–1797). It is known as *Fagnano's Problem*. Information about it can be found on the internet.]

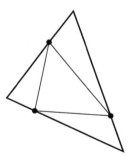

b)   Show that this inscribed triangle is the path of a ball bouncing inside the triangle caught in an infinite loop.

c)   Does an obtuse triangle have an inscribed triangle of shortest perimeter?
See [TANTON1], for instance, for some help on this.

References

[De MARCO] De Marco, L., The conformal geometry of billiards, *Current Events Bulletin*, Joint Mathematics Meetings 2010.

[LAPPAN] Lappan, G., et al. Comparing and scaling: ratio, proportion, and percent, *Connected Mathematics Project*, Prentice Hall, 2004.

[NCTM] National Council of Teachers of Mathematics, Paper pool: analyzing numeric and geometric patterns. URL: http://illuminations.nctm.org/LessonDetail.aspx?ID=U125

[STEINHAUS] Steinhaus, H., *Mathematical Snapshots*, 3rd edition, Dover, New York, NY, 1999.

[TANTON1] Tanton, J., A dozen questions about a triangle, *Math Horizons,* April 2002, 23–28.

[TANTON2] Tanton, J., *THINKING MATHEMATICS! Volume 4: Lines, Circles, Trigonometry and Conics,* www.lulu.com, 2009.

20

Repunits and Primes

PUZZLER: Prime Repunits

A *repunit* is a number all of whose digits are one:

$$1, 11, 111, 1111, 11111, \ldots.$$

Some repunits are prime (such as 11 and 1111111111111111111) and others are composite. (What are the factors of 111, of 1111 and of 11111?)

It is generally believed, but not proven, that infinitely many repunits are prime. (Only five prime repunits are known: those with 2, 19, 23, 317 and 1031 digits.) Your challenge is to establish that

If a repunit is prime, then the number of its digits is prime.

The Primes

A number is *prime* if it has exactly two distinct factors (whole number, positive factors, that is). The list of prime numbers begins

$$2, 3, 5, 7, 11, 13, 17, 19, 23, 29, 31, \ldots.$$

Two consecutive odd numbers that are both prime are called *twin primes*. The list of twin primes begins

$$3\text{-}5, 5\text{-}7, 11\text{-}13, 17\text{-}19, 29\text{-}31, 41\text{-}43, \ldots.$$

Three consecutive odd numbers, each prime, is a called a *triple of primes*. The first triple of primes is 3-5-7. There are no quadruples of primes. (Why?)

The largest known prime, as of the original writing of this newsletter, is $2^{43,112,609} - 1$ with 12,978,189 digits (discovered in August, 2008), the largest known twin primes are

$$65516468355 \cdot 2^{333333} - 1$$

and

$$65516468355 \cdot 2^{333333} + 1$$

each 100,355 digits long (discovered in August 2009) and the largest known triple of primes is 3-5-7.

Is this it? Is there a single largest prime number, or does the list of prime numbers go on forever with ever larger primes for us to discover? Is there a largest pair of twin primes or does the list of twins go on forever too? Is there a largest triple of primes, or is the list of examples infinitely long? (We have presented only one example!)

I know the answer to two of these questions.

The List of Prime Numbers Goes on Forever. The Greek geometer Euclid (ca. 300 B.C.E.) developed an ingenious proof that demonstrates the infinitude of the primes. He gave a method that allows us to find one more prime from those we know. For example, suppose we currently have the primes 2, 3, 5, 7, 11, 13, and 17 and we can't think of any more. Euclid said to multiply them together and to add one.

$$2 \cdot 3 \cdot 5 \cdot 7 \cdot 11 \cdot 13 \cdot 17 + 1 = 510511$$

This is not divisible by 2 (it leaves a remainder of 1), nor is it divisible by 3 (it leaves a remainder of 1), or 5 or 7 or any of the primes we have. So, when we factor 510511 into primes, the primes we see can't be any of the ones we know. Thus we must be seeing new primes. ($510511 = 19 \times 97 \times 277$ giving us the new primes 19, 97, and 277.) We can repeat this process over and over, so the list of primes must be infinite.

The List of Triples of Primes Does NOT Go on Forever. A triple of primes consists of three numbers of the form

$$n \quad n+2 \quad n+4$$

with n odd and each of the three numbers prime. Let's look at remainders on division by three. There are three possibilities:

1. *n is a multiple of three.* Then n won't be prime, unless n is the number three itself. (In which case we discover the triple 3-5-7.)

2. *n is one more than a multiple of three.* Then $n + 2$ will be two more than one more than a multiple of three. That is, $n + 2$ is divisible by three and won't be prime. We have no triples of primes in this case.

3. *n is two more than a multiple of three.* Then $n + 4$ will be divisible by three and so not prime. There are no triples of primes in this case.

Thus the first is the only case that allows for triples of primes, and gives only one example, 3-5-7.

No One Knows Whether or not the List of Twin Primes Goes on Forever. (Or, if they do, they are currently not telling!)

There are infinitely many single primes, and there are only finitely many triple of primes. The twin primes lie between the two. It is curious that we still know nothing about how many examples of them there are!

Prime Formulas

Swiss mathematician Leonhard Euler (1707–1783) discovered the delightful fact that the formula $n(n+1) + 41$ gives a prime number output for each of the forty consecutive inputs $n = 0, 1, 2, 3, \ldots, 39$. (Why doesn't $n = 40$ give a prime output?)

Mathematicians have proven that no simple polynomial formula will ever produce prime outputs for all integer inputs. But they have been searching, nonetheless, for formulas of other types that generate only prime number answers. No significant success has been made in this regard! In 1947 W. H. Mills proved that there exists a number a so that a^{3^n} when rounded down to the nearest integer is always a prime number, but he couldn't say what that number a is. In 1972 H.B. Mann and D. Shanks proved the following bizarre connection to Pascal's triangle:

	0	1	2	3	4	5	6	7	8	9	10	11	12	13	14	15
0	1	★	★	★	★	★	★	★	★	★	★	★	★	★	★	★
1	★	★	1	1	★	★	★	★	★	★	★	★	★	★	★	★
2	★	★	★	★	1	2	1	★	★	★	★	★	★	★	★	★
3	★	★	★	★	★	★	1	3	3	1	★	★	★	★	★	★
4	★	★	★	★	★	★	★	★	1	4	6	4	1	★	★	★
5	★	★	★	★	★	★	★	★	★	★	1	5	10	10	5	1
6	★	★	★	★	★	★	★	★	★	★	★	★	1	6	15	20

Write the rows of Pascal's triangle with row n starting at column 2n. Then a column number is prime precisely when the entries of that column are each divisible by their row number! (For example, column 13 has entries 10, divisible by 5, and 6, divisible by 6, and so 13 is prime. Column 12 has entries that are not divisible by their row numbers.) It's a nifty and surprising result, but it is not a formula that allows us to find prime numbers with ease.

Research Corner. No one currently knows the answers to any of the following challenges. Go for it!

a) Are there infinitely many primes one greater than a square number? (For example, the prime 17 is one greater than 4^2.)

b) How many of the Fibonacci numbers $1, 1, 2, 3, 5, 8, 13, 21, 34, 55, 89, 144, \ldots$ are prime? Finitely many or infinitely many?

c) How many numbers of the form $n^n + 1$ are prime? Currently only $1^1 + 1 = 2$, $2^2 + 1 = 5$ and $4^4 + 1 = 257$ are known.

d) How many numbers of the form $n! + 1$ are prime?

e) How many numbers of the form $2^n - 1$ are prime? (The largest prime known today is of this form.)

f) Is every even number greater than two the sum of two primes?

g) Find a simple formula for the nth prime number.

COMMENTARY, SOLUTIONS and THOUGHTS

The term "repunit" was coined by Albert Beiler in 1966 ([BEILER]) and they appear in the study of periodicity in the decimal expansions of fractions. (For example, it is known that if p is a prime greater than 5, then the period of $\frac{1}{p}$ is the length of the smallest repunit that is divisible by p.) It is standard to denote a repunit of length n as R_n.

If $a \mid n$, then $R_a \mid R_n$. (We have $R_9 = 111, 111, 111 = R_3 \times 1001001$ and $R_{10} = R_2 \times 101010101$, for instance.) Thus

R_n is composite if n is.

As mentioned in the newsletter only five repunits, necessarily of prime length, are known to be prime ([DUBNER and WILLIAMS]) though R_{49081}, R_{86453}, R_{109297} and R_{270343} are likely to also be prime ([WEISSTEIN2]). It is unknown whether the total count of prime repunits is be finite or infinite.

Comment.   There is only one prime repunit in base four, namely 5 (which is 11_4). To see this, note that the base-four repunits are the numbers

$$4^{n-1} + 4^{n-2} + \cdots + 4 + 1 = \frac{4^n - 1}{4 - 1} = \frac{(2^n - 1)(2^n + 1)}{3}.$$

For each n, exactly one of $2^n \pm 1$ is divisible by 3, leaving the other as a proper factor of the repunit (except for the case $n = 2$ where $\frac{2^n - 1}{3} = 1$ and $2^n + 1 = 5$ are improper factors).

Comment.   The Mersenne primes are repunits in base two with prime counts of digits.

The number 31 is a repunit in two different bases (it is 111_5 in base 5 and 11111_2 in base 2), as is the number 8191 (which is 111_{90} in base 90 and $111,111,111,111_2$ in base 2). The *Goormaghtigh Conjecture* states that these are the only two (non-trivial) integers that are repunits in at least two different bases. (See [GUY].)

Repdigits and Squares

I sent out the following puzzler to the e-mail followers of the Math Institute:

> **Square Repdigits.** *A "repdigit" is a counting number all of whose digits are the same. For example, 44444 is a repdigit, as are 55, 7777777777777777777777, and 8.*
>
> *There are only finitely many repdigits that are square numbers. Find the largest square repdigit you can and try to prove that it is the largest.*

Experimentation leads us to suspect that 1, 4 and 9 are the only square repdigits. This is the case as the following series of observations show.

Observation 1.   *The final digit of a square number can only be 0, 1, 4, 5, 6, or 9.*
Consequently there no square repdigits using the digits 2, 3, 7, or 8.

Observation 2.   *If a square n^2 is divisible by a prime p, it is divisible by p^2.*
(This follows from unique factorization.)
Consequently there is no square of the form $555 \cdots 55 = 5 \times 111 \cdots 11$ or of the form $666 \cdots 66 = 2 \times 3 \times 111 \cdots 11$.

Observation 3.   *No repunit (other than 1) is square.*

If $111 \cdots 11 = n^2$, then n must be an integer with final digit 1 or 9. Thus $n = 10k \pm 1$. Consequently, $n^2 = 100k^2 \pm 20k + 1$ has second-to-last digit even. This is not the case for a repunit other than the repunit 1. (Alternatively, consider the repunits modulo 4.)

Observation 4. *If* $444 \cdots 44 = 4 \times 111 \cdots 11$ *or* $999 \cdots 99 = 9 \times 111 \cdots 11$ *is square, then so is a repunit.*

Thus 1, 4 and 9 are the only possible square repunits.

Comment. John McCarthy, an avid follower of the Math Institute activities (and also an alumnus of St. Mark's School) showed that squares, and even fourth powers, occur as repdigits in other bases. For example, he found:

$$1111_7 = 20^2 \qquad\qquad 55_{19} = 10^2 \qquad\qquad AA_{342} = 294^2 (A = 252)$$

$$777_{18} = 49^2 = 7^4 \qquad BB_{47} = 6^4 (B = 27) \qquad CC_{342} = 14^4 \ (C = 112)$$

Polynomials and Primes

Euler's wonderful polynomial $p(n) = n^2 + n + 41$ gives prime outputs for the eighty integer inputs $n = -40, -39, \ldots, 38, 39$. (It gives 40 distinct prime outputs for $n = 0, \ldots, 39$.) However $p(40) = 40 \cdot 41 + 41$ is divisible by 41.

It is not difficult to prove that there can be no polynomial $p(n)$ with integer coefficients (of positive degree) such that $p(1), \ p(2), \ p(3), \ldots$ are prime. Actually, there is no integer polynomial with $p(n)$ prime for all $n > N$ for some N ("eventually always prime").

Suppose $p(n)$ is such a polynomial. We can assume that its leading coefficient is greater than or equal to 1 so that $p(n) \to \infty$ as $n \to \infty$. Moreover, we can assume that there is a value N so that $p(n)$ is strictly increasing for $n \geq N$. And by choosing N we can also assume that all the outputs $p(n)$ are prime in this range.

So, if $q = p(N)$ (a prime), then

$$p(N), \ p(N+q), \ p(N+2q), \ p(N+3q), \ \ldots$$

is an increasing sequence of primes.

The binomial theorem shows that $p(x + y)$, when expanded, possesses terms with factor y plus a collection of terms that match $p(x)$, so $p(x + y) = p(x) + y(\cdots)$. Consequently

$$p(N + kq) = p(N) + q(\cdots) = q + q(\cdots)$$

and so is divisible by q. This contradicts the assumption that our sequence consists only of primes.

Comment. Christian Goldbach proved that

There is no non-constant polynomial with real coefficients that is always prime for positive integer inputs.

See [YOUNG, page 64] for details and proof.

Although polynomials produce infinitely many composite outputs for integer inputs, there is a host of charmers that, like Euler's polynomial, give us a run for our money! For example:

$2n^2 + 29$ is prime for $n = 0, 1, \ldots, 28$ (Legendre, 1798)

$n^2 + n + 17$ is prime for $n = 0, 1, \ldots, 15$ (Legendre, 1798)

$6n^2 - 342n + 4903$ is prime for $n = 0, 1, \ldots, 57$ (Brox, 2006)

$n^4 - 97n^3 + 3294n^2 - 45458n + 213589$ is prime for $n = 0, 1, \ldots, 49$ (Beyleveld, 2006)

(See [WEISSTEIN1] for a compilation of such curiosities.)

For a collection of unusual prime-generating algorithms see [DUDLEY].

Challenge. Prove that $n^2 + n + 41$ is never divisible by a prime smaller than 41.

References

[BEILER] Beiler, A., *Recreation in the Theory of Numbers: The Queen of Mathematics Entertains*, Dover, New York, NY, 1966.

[DUBNER and WILLIAMS] Dubner, H., and Williams, H. C., The Primality of R_{1031}, *Mathematics of Computation,* **47** (1986), 703–711.

[DUDLEY] Dudley, U., Formulas for primes, *MathematicsMagazine*, **56** (1983), 17–22.

[GUY] Guy, R., *Unsolved Problems in Number Theory*, 3rd ed., Springer, New York, 2004.

[WEISSTEIN1] Weisstein, E., Prime-generating polynomial. From *MathWorld*—AWolfram Web Resource: URL: http://mathworld.wolfram.com/Prime-GeneratingPolynomial.html

[WEISSTEIN2] Weisstein, E., Repunit. From *MathWorld*—A Wolfram Web Resource. URL: http://mathworld.wolfram.com/Repunit.html

[YOUNG] Young, R., *Excursions in Calculus: An Interplay of the Continuous and the Discrete,* The Mathematical Association of America, Washington D.C., 1992.

21

The Stern-Brocot Tree

PUZZLER: A Curious Fraction Tree

Here is something fun to think about. Consider the fraction tree:

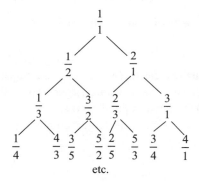

etc.

It is constructed by the rule

$$\frac{a}{b}$$

$$\frac{a}{a+b} \qquad \frac{a+b}{b}$$

Each fraction has two children: a left child, a fraction smaller than 1, and a right child larger than 1.

a)   Continue drawing the fraction tree for another two rows.

b)   Explain why the fraction $\frac{13}{20}$ will appear in the tree. (It might be easier to first figure what should be the parent of $\frac{13}{20}$, its grandparent, and so on.)

c)   Might the fraction $\frac{13}{20}$ appear twice in the tree?

d)   Will the fraction $\frac{457}{777}$ appear in the tree? Might it appear twice?

e)   Are all the fractions that appear in the tree in reduced form? (I see 2/3, but I don't see its equivalents 4/6 or 12/18.)

f)   Let me give things away: Prove that the tree produces *all* possible positive fractions, written in reduced form, with no fraction appearing more than once.

The Euclidean Algorithm

If two numbers a and b are multiples of three, then so too will be their difference: $b - a$. In fact, it is not too hard to see that if we list the common factors of a and b and then list the common factors of a and $b - a$, the two lists will be the same. [For example, with $a = 84$ and $b = 120$ the pair $(84, 120)$ has common factors: 1, 2, 3, 4, 6, and 12, and the pair $(84, 36)$ has common factors: 1, 2, 3, 4, 6, and 12.]

The Greek geometer Euclid (ca. 300 B.C.E) noticed this too and went further to observe that if we repeatedly subtract the smaller number from the larger from a pair of numbers to produce new pairs, the lists of common factors cannot change:

$$(84, 120) \rightarrow (84, 36) \rightarrow (48, 36)$$
$$\rightarrow (12, 36) \rightarrow (12, 24) \rightarrow (12, 12)$$

All pairs have common factors

1, 2, 3, 4, 6 and 12.

As the numbers in the pairs are decreasing (we do not allow ourselves to go to zero), the process stops when there is no smaller value to subtract from a larger. That is, the process stops by reaching a pair with a common value: (d, d). The pair of numbers d and d has the same list of common factors as the original pair a and b. Since d is clearly the largest common factor of d and d, d is the largest common factor of a and b.

Euclidean Algorithm. *To find the greatest common factor of two numbers, repeatedly subtract the smaller from the larger until a common value appears. This common value is the greatest common factor!*

For example, consider 102 and 170. The algorithm gives:

$$(102, 170) \rightarrow (102, 68) \rightarrow (34, 68) \rightarrow (34, 34)$$

Their greatest common factor is 34.

Exercise. Find the greatest common factor of 457 and 777.

Connections to the Fraction Tree. Start at apex $\frac{1}{1}$ of the fraction tree and follow a path down to some fraction $\frac{a}{b}$. Do you see that the path is just the Euclidean algorithm in reverse? Starting at a/b and following the path back to $1/1$ is precisely the path of Euclid's algorithm applied to the pair (a, b). Because the path yields the final pair $(1, 1)$, the fraction a/b has greatest common factor 1 between its numerator and denominator. That is, a/b is in reduced form. Moreover, every reduced fraction must appear in the tree: apply the Euclidean algorithm to the numerator and denominator to see where it lies in the tree.

No fraction appears twice in the tree. If one did, then its parent would appear twice, as would its grandparent, great-grandparent, and so on, all the way down to 1/1 appearing twice, which it does not!

Comment. If we read the fractions left to right across the rows of the tree we obtain the sequence: $\frac{1}{1}, \frac{1}{2}, \frac{2}{1}, \frac{1}{3}, \frac{3}{2}, \frac{2}{3}, \frac{3}{1}, \frac{1}{4}, \frac{4}{3}, \ldots$. This list contains <u>all</u> the positive fractions. The fact that one can list the rationals was first discovered by Georg Cantor in the 1800s. He accomplished this in a different way.

Challenge. Put <u>all</u> the rationals, both the positive and negative fractions, including zero, into a single list.

Some Mysteries. Look at the numerators that appear in the fraction tree, reading across the tree by rows. We obtain the sequence

$$1\ 1\ 2\ 1\ 3\ 2\ 3\ 1\ 4\ 3\ 5\ 2\ 5\ 3\ 4\ 1\ 5 \ldots$$

(The denominators follow the same sequence offset by one.)

a)   Consider every second entry, the ones in even positions. This gives 1, 1, 2, 1, 3, 2, 3, 1, ..., the original sequence!

b)   Each entry in an odd position (after the first) is the sum of its two neighbors!

$$2 = 1 + 1, \ 3 = 1 + 2, \ 3 = 2 + 1, \ 4 = 1 + 3, \ 5 = 3 + 2, \ldots$$

c)   Every third entry is even. All other entries are odd.

d)   The entries between the ones are palindromes: 2   323   4352534   547385727583745. . . . The digits of each sum to one less than a power of three.

Research Corner.

i)   Prove the claims above.

ii) Discover more remarkable properties of this sequence.

iii) Push the sequence backwards. For instance, b) says that the entry to the left of the first 1 must be zero. What's next on the left?

COMMENTARY, SOLUTIONS and THOUGHTS

The remarkable fraction tree shown below is known as the Stern-Brocot tree and it appears in [GRAHAM, KNUTH and PATASHNIK]. A similar construct, known as the Calkin-Wilf tree, is given by a similar algorithm: each reduced fraction $\frac{a}{b}$ has left child $\frac{a}{a+b}$ but right child $\frac{b}{a+b}$. Its properties, like those we have outlined in the newsletter, are presented in [CALKIN and WILF]. The content of the newsletter also appears in [TANTON].

In the fall of 2009, young students of the St. Mark's Institute of Mathematics research class explored the properties of the Stern-Brocot tree.

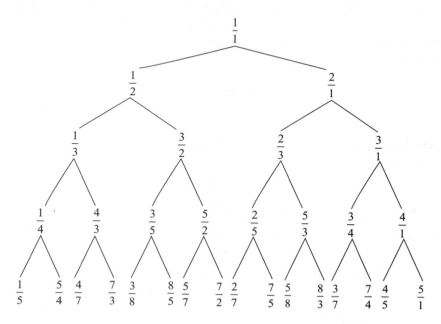

They asked the questions

If we enumerate the fractions reading across the rows of the tree we obtain a sequence of all positive rationals

$$\frac{1}{1}, \frac{1}{2}, \frac{2}{1}, \frac{1}{3}, \frac{3}{2}, \frac{2}{3}, \frac{3}{1}, \frac{1}{4}, \frac{4}{3}, \frac{3}{5}, \frac{5}{2}, \frac{2}{5}, \frac{5}{3}, \frac{3}{4}, \frac{4}{1}, \frac{1}{5}, \frac{5}{4}, \frac{4}{7}, \ldots :$$

what is the 100th fraction in this list? The 400,003rd fraction? Is it possible to determine the N th fraction without having to enumerate entire rows of this tree?

Here is their result:

Theorem 1. *Write N in base two and count the digits that appear in each block of 1s and 0s in its representation. Suppose*

$$N = \overbrace{1 \cdots 1}^{a_k} \overbrace{0 \cdots 0}^{a_{k-1}} \overbrace{1 \cdots 1}^{a_{k-2}} \cdots \overbrace{0 \cdots 0}^{a_1} \overbrace{1 \cdots 1}^{a_0}$$

(with $a_0 = 0$ if N is even). Then the Nth fraction f_N in the Stern-Brocot sequence is the continued fraction

$$f_N = [a_0, a_1, a_2, \ldots, a_k] = a_0 + \cfrac{1}{a_1 + \cfrac{1}{a_2 + \cdots \cfrac{1}{a_k}}} .$$

For example, the number 100 in base two is 1100100 and so

$$f_{100} = 0 + \cfrac{1}{2 + \cfrac{1}{1 + \cfrac{1}{2 + \cfrac{1}{2}}}} = \frac{7}{19}$$

and 400003 in base two is 1100001101010000011 and so

$$f_{400003} = [2, 5, 1, 1, 1, 1, 2, 4, 2] = \frac{1553}{713}.$$

Proof. Construct a tree of similar shape using the counting numbers

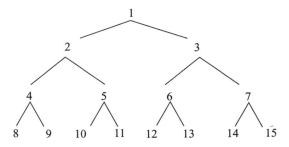

It follows the rule

For numbers written in binary, this has a nice interpretation:

To create the left child of a counting number, append a 0 at the end of its binary representation. To create a right child, append a 1.

Thus each digit 0 in a binary representation of a number represents a left down-step in the tree and each 1 (after the first) a right down-step in the tree. The number fourteen, for example, with binary code 1110 can be interpreted as "1RRL" which means:

Start at 1. Step right. Step right. Step left.

In terms of blocks, if $N = \overbrace{1 \cdots 1}^{a_k} \overbrace{0 \cdots 0}^{a_{k-1}} \overbrace{1 \cdots 1}^{a_{k-2}} \cdots \overbrace{0 \cdots 0}^{a_1} \overbrace{1 \cdots 1}^{a_0}$ has blocks $a_k, a_{k-1}, \ldots, a_1, a_0$ (with a_0 possibly zero), then its left child, addend a zero, has blocks $a_k, a_{k-1}, \ldots, a_1, a_0, 1, 0$ if $a_0 \neq 0$ and blocks $a_k, a_{k-1}, \ldots, a_1 + 1, 0$ if $a_0 = 0$, and its right child, addend a 1, has blocks $a_k, a_{k-1}, \ldots, a_1, a_0 + 1$.

The fractions in the Stern-Brocot tree follow the same pattern! If

$$\frac{a}{b} = a_0 + \cfrac{1}{a_1 + \cfrac{1}{a_2 + \cdots + \cfrac{1}{a_k}}}$$

then its left child is:

$$\frac{a}{a+b} = \cfrac{1}{1+\cfrac{1}{a/b}} = 0 + \cfrac{1}{1+\cfrac{1}{a_0+\cfrac{1}{a_1+\cfrac{1}{a_2+\cdots+\cfrac{1}{a_k}}}}}$$

(which equals $\dfrac{a}{a+b} = \cfrac{1}{1+\cfrac{1}{a/b}} = 0 + \cfrac{1}{a_1+1+\cfrac{1}{a_2+\cdots+\cfrac{1}{a_k}}}$ if $a_0 = 0$), and its right child is

$$\frac{a+b}{b} = \frac{a}{b}+1 = (a_0+1)+\cfrac{1}{a_1+\cfrac{1}{a_2+\cdots+\cfrac{1}{a_k}}}.$$

This establishes the proof.   □

Students went further and found the following curiosity:

Theorem 2. *For* $\frac{1}{1}, \frac{1}{2}, \frac{2}{1}, \frac{1}{3}, \frac{3}{2}, \frac{2}{3}, \frac{3}{1}, \frac{1}{4}, \frac{4}{3}, \frac{3}{5}, \frac{5}{2}, \frac{2}{5}, \frac{5}{3}, \frac{3}{4}, \frac{4}{1}, \frac{1}{5}, \frac{5}{4}, \frac{4}{7}, \ldots$, *the sum of any fraction in the sequence and the reciprocal of its next term is always an odd integer:* $f_{N-1} + \frac{1}{f_N} = 2a_0 + 1$ *where* a_0 *is the number of zeros in the tail of the binary representation of* N.

This provides a quick means of computing the Nth fraction f_N if f_{N-1} is known.

The Numerators in the Fraction Tree

Let f_N denote the Nth fraction in the Stern-Brocot tree in reading left to right across its rows. It is not difficult to see that the denominator of f_N always matches the numerator of f_{N+1}. Thus if e_N denotes the numerator of f_N we have:

$$f_N = \frac{e_N}{e_{N+1}}$$

with the sequence of numerators beginning 1, 1, 2, 1, 3, 2, 3, 1, ...
From

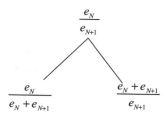

we see $e_{2N} = e_N$ and $e_{2N+1} = e_N + e_{N+1}$. This shows that the even terms in the sequence $\{e_N\}$ match the original sequence and that the odd terms satisfy $e_{2N+1} = e_N + e_{N+1} = e_{2N} + e_{2N+2}$, as claimed in the newsletter. The palindromic properties of the sequence follow as a consequence.

(The sequence starts with the palindromes 1 and 121 and these represent the next few even terms of the sequence perpetuating the symmetry. The odd terms, being the sums of neighboring terms, preserve it too.)

Another Binary Connection

Every positive integer has a unique binary representation using the digits 0 and 1. If we permit use of the digit 2 as well, a number may be expressed in base two in more than one way. For example, the number ten now has five possible representations:

$$1010 \leftrightarrow 8 + 2$$
$$1002 \leftrightarrow 8 + 1 + 1$$
$$210 \leftrightarrow 4 + 4 + 2$$
$$202 \leftrightarrow 4 + 4 + 1 + 1$$
$$122 \leftrightarrow 4 + 2 + 2 + 1 + 1$$

I present this to students as a weighing puzzle:

> *I have two rocks each weighing 1 pound, two rocks each weighing 2 pounds, two each weighing 4 pounds, two each weighing 8 pounds, and so on. In how many ways can I make ten pounds?*

Let $b(N)$ denote the count of ways of making a weight of N pounds. Then $\{b(N)\}$ is known as *Stern's diatomic sequence* [SLOANE, A002487].

If N is odd, then we must use precisely one 1-pound rock and we're left trying to make a weight of $N - 1$ pounds using rocks of weights 2, 4, 8, 16 .... Dividing by two we see that this is equivalent to computing $b\left(\frac{N-1}{2}\right)$. Thus we have

$$b(2k + 1) = b(k).$$

If N is even, we can either use two 1-pound rocks and match a weight of $N - 2$ pounds with the remaining rocks (there are $b\left(\frac{N-2}{2}\right)$ ways to do this) or we can use no 1-pound rocks and match N pounds with the remaining rocks (there are $b\left(\frac{N}{2}\right)$ ways to accomplish this). This gives

$$b(2k) = b(k) + b(k + 1).$$

These relations allow us to write down the sequence $\{b(N)\}$:

$$1, 2, 1, 3, 2, 3, 1, 4, 3, 5, 2, 5, 3, 4, 1, \ldots$$

and we see the sequence $\{e_N\}$ offset by one! (The recursive relations for e_{N-1} and $b(N)$ are the same.)

Given our work with the Stern-Brocot tree, the following result is clear. But if it were presented cold, with no context for its derivation, it is truly mind boggling!

Theorem 3. *Consider the sequence* $1, 2, 1, 3, 2, 3, 1, 4, 3, 5, 2, 5, 3, 4, 1, \ldots$ *given by counting the number of ways to represent each integer in base two using the digits 0, 1, and 2. Then every pair of coprime integers (a, b) appears in the sequence as a pair of neighboring terms exactly once.*

We can say more about this sequence by rephrasing Theorem 2:

> *For any three consecutive terms in the sequence, the sum of the first and last terms is an odd multiple of the middle term.*

References

[CALKIN and WILF] Calkin, N., and Wilf, H.S., Recounting the rationals, *American Mathematical Monthly,* **107** (2000), 360–363.

[GRAHAM, KNUTH and PATASHNIK] Graham, R.L., Knuth, D.E., and Patashnik, O., *Concrete Mathematics: A Foundation for Computer Science, 2nd ed.*, Addison-Wesley, Reading, MA, 1994.

[SLOANE] Sloane, N.J.A., The On-Line Encyclopedia of Integer Sequences. URL: www.research .att.com/~njas/sequences

[TANTON] Tanton, J., *THINKING MATHEMATICS! Volume 1:Arithmetic = Gateway to All*, www.lulu.com, 2009.

22 Tessellations

PUZZLER: Tessellations and Tilings

A polygon is said to *tessellate* the plane if it is possible to cover the entire plane with congruent copies of it without overlap (except along the edges of the figures). The tessellation is called a *tiling* if each edge of one polygon matches an entire edge of an adjacent polygon.

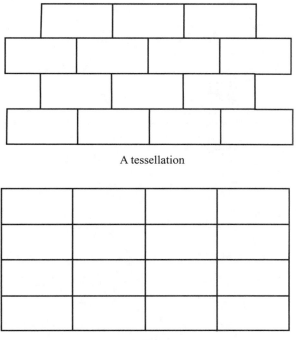

A tessellation

A tiling

Parallelograms tile the plane. As two copies of the same triangle can be placed side-by-side to form a parallelogram we see that *every triangle tiles the plane*.

1. The tiling with triangles shown is *periodic*, meaning that it possesses translational symmetries in (at least) two non-parallel directions. That is, it is possible to shift the entire tiling along some direction of translation and return to precisely the same covering of the plane, and again along some second, non-parallel direction. Is it possible to use a triangular tile to create a non-periodic tiling of the plane?

[Comment. The definition of "periodic" requires a tiling to possess at least two independent translational symmetries. Is it possible for a tiling to possess translational symmetry in just *one* direction? If so, construct an example.]

2. Squares, rectangles, and parallelograms all tile the plane. Does every quadrilateral tile the plane? (Even concave quadrilaterals?)

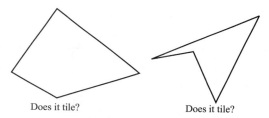

Does it tile?   Does it tile?

3. Is there a five-sided polygon that tessellates the plane? How about one that tiles the plane?

Think about these puzzles before reading on.

Does It Tile?

Regular hexagons tile the plane:

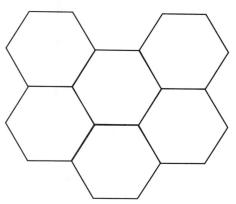

In fact, any hexagon with opposite sides parallel will tile the plane

Two copies of a quadrilateral (convex or concave) placed side-by-side produces a hexagon with parallel opposite sides.

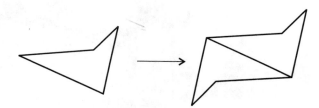

This shows that all quadrilaterals tile the plane:

There is a pentagon that tessellates the plane:

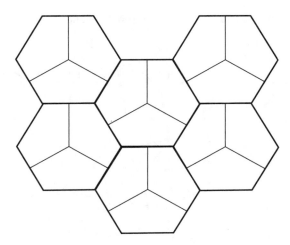

There is even one that tiles the plane! Remove an equilateral triangle from a square to produce a five-sided shape. It tiles!

In summary,

> *Every triangle tiles the plane.*
> *Every quadrilateral tiles the plane.*
> *There exist pentagons that tile the plane.*
> *There exist hexagons that tile the plane.*

Question. Why doesn't a regular pentagon tile the plane? (Which regular polygons do?) Draw an example of a hexagon that does not tile the plane.

For each $n \geq 7$ it is easy to create a *concave* n-gon that tiles the plane.

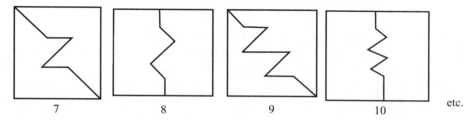

Do there exist *convex* n-gons, with $n \geq 7$, that tile the plane? Mathematicians know the answer. It's in the negative.

Theorem. *No convex n-gon with $n \geq 7$ tiles the plane and no convex n-gon with $n \geq 7$ tessellates the plane.*

A Non-Periodic Tiling. An isosceles triangle with appropriate dimensions can be used to make a non-periodic tiling of the plane. The rotational symmetry does not translate.

Split the design to make a spiral!

Research Corner. It remains an unsolved problem as to which pentagons tessellate the plane. Currently only 14 types of pentagons are known to work. Care to finish solving this classification problem? [High-schoolers in New South Wales, Australia, have contributed to this project!]

COMMENTARY, SOLUTIONS and THOUGHTS

In the newsletter we presented a result due to Ivan Niven:

Theorem. *No convex n-gon with $n \geq 7$ tessellates the plane.*

Here is an overview of the proof. The full details can be found in [NIVEN1] (and also [NIVEN2] for the case of a tiling).

Proof. Suppose we have a tessellation of the plane with a convex n-gon, $n \geq 7$. Suppose the plane has coordinate axes scaled so that the perimeter of each tile is one unit in length. (Each tile consequently has width no more than half a unit.) Let A be the area of each tile.

For each real number r let S_r be the set of tiles in the tessellation that each cover some point in the disc of radius r centered about the origin, and let $|S_r|$ denote the number of tiles in it. As no tile has width greater than $\frac{1}{2}$, the tiles of S_r cover the disc of radius r but do not extend beyond a disc of radius $r + 1$. Consequently

$$\pi r^2 \leq |S_r| \cdot A < \pi (r + 1)^2 .$$

Let v be the number of vertices that appear in the sub-tessellation given by S_r. Of the 2π radians of turning that surrounds each vertex, some portion corresponds to the measure of interior angles of tiles. (In a tiling, all interior vertices of the sub-tiling are fully surrounded by interior angles of tiles, and boundary vertices only partially so. For a tessellation, some interior vertices lie on the side of another tile and are not fully surrounded by tile interior angles.) The quantity $2\pi v$ is an over estimate of the sum of the interior angles of the tiles of S_r. As the sum of interior angles of one tile is $(n - 2)\pi$ we have

$$2\pi v \geq (n - 2)\pi \cdot |S_r|.$$

On the other hand, since each tile is convex, at least three tiles surround each vertex in the full tessellation. As all the vertices that appear in S_r are interior vertices of the sub-tessellation S_{r+1}, and as each tile of S_{r+1} has n vertices we have

$$n|S_{r+1}| \geq 3v.$$

These sets of inequalities give a contradiction. The last two inequalities give

$$\frac{n}{3}|S_{r+1}| \geq \frac{n - 2}{2}|S_r|.$$

Multiplying by A and using the first inequality gives

$$\frac{n}{3}\pi (r + 2)^2 > \frac{n}{3}|S_{r+1}| \cdot A \geq \frac{n - 2}{2}|S_r| \cdot A \geq \frac{n - 2}{2}\pi r^2.$$

Consequently

$$\frac{2n}{3n - 6} \geq \left(\frac{r}{r + 2}\right)^2 .$$

Algebra shows that for $n \geq 7$ we have $\frac{2n}{3n-6} \leq \frac{14}{15}$, and so

$$\left(\frac{r}{r + 2}\right)^2 < \frac{14}{15}$$

for all r. This is a contradiction as the left-hand side approaches the value 1 as r increases.   □

Non-Periodic Tessellations and Tilings

Tile the plane with squares and divide each tile along its diagonal to obtain a tiling with isosceles right triangles. If we do this at random there is no translational symmetry and we have a non-periodic tiling of the plane.

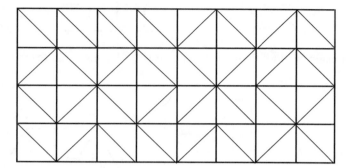

Comment. In a sense these directions are unsatisfactory: How can we specify the direction of each diagonal for an infinite array of cells and be sure that no translational symmetry holds? Does choosing a diagonal at random an infinite number of times define a tiling? It would be better to specify an algorithm for assigning choices, perhaps based on the distribution of prime numbers, for instance—one that we can prove disrupts periodicity. Care to try?

We can use this approach to produce a tiling of the plane with isosceles right triangles possessing only *one* direction of translational symmetry: declare one column of squares to be "column 1" and assign one type of square diagonal to the prime-numbered columns to its right.

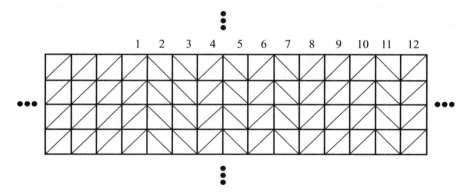

Challenge. Prove that if p is prime, there is no number k such that $p, p + k, p + 2k, p + 3k, \ldots$ are all prime.

A tile is said to *self-replicate* if a finite number of congruent copies of it fit together to make a larger copy of the tile. Self-replicating tiles provide an interesting source of tessellations for the plane. For example, four copies of a square stack together to make a larger square. Repeating this produces the regular square tiling of the plane:

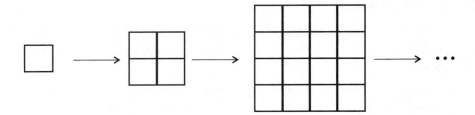

All triangles self-replicate, as does the L-shape and an isosceles trapezoid (each in more than one way!)

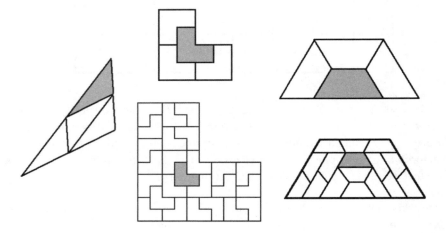

Challenge. Find a figure with the property that two copies of it fit together to produce a scaled version of that figure.

Challenge. The L-shape shown above is subdivided into 4 and 25 copies of itself. Show that we can also subdivide the L-shape into 9 and 16 copies of itself. Can it be subdivided into a non-square number of copies?

I like the 1-2-$\sqrt{5}$ right triangle: five copies of it stack together to form another 1-2-$\sqrt{5}$ right triangle.

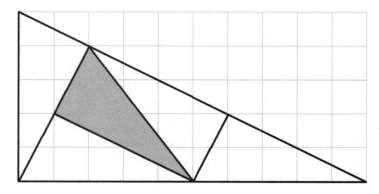

If we iterate an infinite number of times we obtain a tessellation of the entire plane, called the *pinwheel tessellation*. (See [RADIN].)

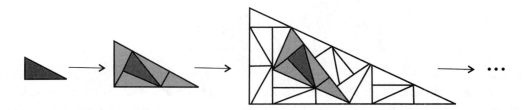

The tessellation is non-periodic!

Comment. Students at St. Mark's School have covered a corridor wall with six iterations of the self-replication starting with 12-inch wide construction-paper triangles. The resultant, and enormous, product is mesmerizing! The tiling was also used to decorate the Federation Square Buildings in Melbourne, Australia. (See, for example, [WEISSTEIN].)

The non-periodicity of the tessellation follows from the following lemma. (Unfortunately, proving the lemma takes some work! We'll leave this to the end of our discussion.)

Lemma. *The smallest angle in a 1-2-$\sqrt{5}$ right triangle, namely $x = \arctan\left(\frac{1}{2}\right)$, is an irrational multiple of 2π.*

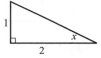

The hypotenuse of the initial right triangle makes an angle x with the horizontal. We depict this as

The first iteration of self replication produces a line segment of angle x with the horizontal (its hypotenuse) and a central line segment of angle $2x$ to the horizontal (the hypotenuse of the central copy of the original):

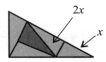

And the second iteration produces line segments of angles x, $2x$ and $3x$ to the horizontal:

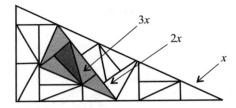

And so on.

As x is an irrational multiple of 2π, the infinite list of multiples of x contains no repeat angles (even modulo 2π). Thus the infinite tessellation of the plane contains line segments at infinitely many different angles to the horizontal.

If the tessellation is periodic, there must be two directions of translation symmetry. That is, there is a distance a so that a translation through it produces an identical copy, and a distance b in a second direction with the same property. Thus there is a parallelogram defined by these two isometries that offers the template for the repeating design of the tessellation. Because it is bounded it can contain only a finite number of line segment angles, forcing the same for the entire tessellation. As we have seen, this is not the case. Thus the tessellation cannot be periodic.

Hard Challenge. Is the tessellation resulting from the self-replication of the L-shape periodic?

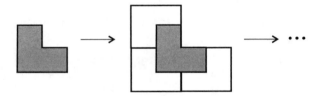

Proof of Lemma. We need to establish that $\arctan\left(\frac{1}{2}\right)$ is an irrational multiple of 2π. The proof here is based on the work of J.M.H. Olmsted ([OLMSTED]).

We begin with some facts about the tangent function. We know that

$$\tan(a+b) = \frac{\tan a + \tan b}{1 - \tan a \cdot \tan b}$$

so if $\tan x = t$, then

$$\tan 2x = \frac{2t}{1-t^2}, \quad \tan 3x = \frac{t^3 - 3t}{3t^2 - 1}, \quad \tan 4x = \frac{4t^3 - 4t}{-t^4 + 6t^2 - 1},$$

and so on. An induction argument establishes that

$$\tan nx = \frac{p(t)}{q(t)}$$

with $p(t)$ and $q(t)$ polynomials in t of the form:

$$p(t) = nt^{n-1} + \cdots \text{ and } q(t) = -t^n + \cdots \text{ if } n \text{ is even,} \tag{1}$$

$$p(t) = t^n + \cdots \text{ and } q(t) = nt^{n-1} + \cdots \text{ if } n \text{ is odd.} \tag{2}$$

This shows

Observation 1. *If* $\tan x = t$ *is rational, then* $\tan nx$ *is also (if it is finite).*

Observation 2. *If* $\tan x = \frac{p}{q}$ *for integers* p *and* q *and* $\tan nx = 0$, *then* $\frac{p}{q}$ *is a rational root of a polynomial with leading coefficient* n *or* 1. *We thus have* $q \mid n$.

We are interested in the case where $\tan x = \frac{1}{2}$. By Observation 1, each value of $\tan nx$ is also rational (when it is finite). Write

$$\tan x = \frac{p_n}{q_n}$$

as a fraction in reduced form (when the quantity is finite).

Observation 3. *The denominators* q_n *can be arbitrarily large.*

To see this, suppose that $\tan \alpha = \frac{p}{q}$ is a fraction in reduced form with one of p or q even, with q is positive. Then it is not difficult to see that $\tan 2\alpha = \frac{2pq}{q^2 - p^2}$ is also in reduced form. Moreover, if $q > |p|$, then $q^2 - p^2 = (q - |p|)(q + |p|) > q + |p| > q$, and if $|p| > q$, then $p^2 - q^2 > |p| + q > q$. Either way, when $\frac{2pq}{q^2 - p^2}$ is written as a fraction with positive denominator the denominator is larger than q.

We have $\tan x = \frac{1}{2}$, which is a fraction in reduced form with even denominator so the denominators of $\tan 2x$, $\tan 4x$, $\tan 8x$, ... grow without bound.

We are now set to complete the proof.

Suppose $\tan x = \frac{1}{2}$ with $x = \frac{a}{b} \cdot 2\pi$ for some integers a and b. Let $\alpha = nx$. Then $\tan \alpha = \tan nx = \frac{p_n}{q_n}$ and $\tan(b\alpha) = \tan(na \cdot 2\pi) = 0$. By Observation 2, this means that $q_n \mid b$ for all n. Thus each value q_n is bounded by b. This contradicts Observation 3. □

References

[NIVEN 1] Niven, I., Convex polygons that cannot tile the plane, *American Mathematical Monthly*, **54** (1978), 785–792.

[NIVEN 2] Niven, I., *Maxima and Minima without Calculus*, Mathematical Association of America, Washington D.C., 1981.

[OLMSTED] Olmsted, J.M.H., Rational values of trigonometric functions, *American Mathematical Monthly*, **52** (1945), 507–508.

[RADIN] Radin, C., The pinwheel tilings of the plane, *Annals of Mathematics*, **139** (1994), 661–702.

[WEISSTEIN] Weisstein, E. W., Aperiodic tiling from *MathWorld*-A Wolfram Web Resource. URL: http://mathworld.wolfram.com/AperiodicTiling.html

23

Theon's Ladder and Squangular Numbers

PUZZLER: Irrational Numbers

Everyone knows that $\sqrt{2}$ is irrational.

a)   Prove it! That is, prove there is something mathematically wrong in writing $\sqrt{2} = a/b$ for some integers a and b.

b)   Prove that $\sqrt{3}$ is irrational.

c)   Prove that $\sqrt{6}$ is irrational.

d)   Prove that $\sqrt{2} + \sqrt{3}$ is irrational.

e)   That $\sqrt{2}$ can be written as $2^{1/2}$ shows that it is possible to raise a rational number to a rational power to obtain an irrational result. Is it possible to raise an irrational number to an irrational power to obtain a rational result?

Squangular Numbers

The square numbers begin 1, 4, 9, 16, 25, ... We can arrange a square number of pebbles into a square array.

The triangle numbers begin: 1, 3, 6, 10, 15, ... We can arrange a triangle number of pebbles into a triangular array.

The number 36 is both square and triangular. I call it a "squangular number."

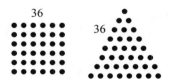

We have that 1 is the first squangular number, and 36 is the second. What's the third squangular number? How many squangular numbers are there?

Read on!

Theon's Ladder. Theon of Smyrna (ca. 140 C.E.) knew that if a/b is a fraction that approximates $\sqrt{2} = 1.414\ldots$, then $\frac{a+2b}{a+b}$ is a better approximation.

For example, $\frac{3}{2} = 1.5$ is close to $\sqrt{2}$, but $\frac{3+4}{3+2} = \frac{7}{5} = 1.4$ is closer still.

Starting with $1/1$ and iterating Theon's method gives a table of values today known as Theon's Ladder:

a	1	3	7	17	41	99	239	577	1393	...
b	1	2	5	12	29	70	169	408	985	...

[Notice: $1393/985 = 1.4142132\cdots$.]

Many wonderful patterns lie within this table!

1. Square the numbers in each row. We see that the top row is alternately one more and one less than double the bottom.

a^2	1	9	49	289	1681	9801	...
b^2	1	4	25	144	841	4900	...

[Algebra proves this. If $a^2 = 2b^2 \pm 1$ then it is easy to check that $(a + 2b)^2 = 2(a + b)^2 \mp 1$.]

This means that as we move along the table the quantity $(\frac{a}{b})^2$ has values closer and closer to 2. Thus Theon's Ladder gives better and better approximations to $\sqrt{2}$.

Challenge. Establish that if a/b on the ladder differs from $\sqrt{2}$ by an amount E, then the next term $(a + 2b)/(a + b)$ differs from $\sqrt{2}$ by an amount less than $E/4$.

2. Look at every odd term of the ladder:

$$\frac{1}{1} = \frac{0+1}{1} \text{ and } 0^2 + 1^2 = 1^2$$

$$\frac{7}{5} = \frac{3+4}{5} \text{ and } 3^2 + 4^2 = 5^2$$

$$\frac{41}{29} = \frac{20+21}{29} \text{ and } 20^2 + 21^2 = 29^2$$

$$\frac{239}{169} = \frac{119+120}{169} \text{ and } 119^2 + 120^2 = 169^2$$

The odd terms of the ladder give Pythagorean triples whose legs differ by one!

Challenge. Prove that the ladder encodes *all* such Pythagorean triples.

3. Look at the even terms of the ladder:

$$\frac{3}{2} = \frac{2(1)+1}{2(1)}$$

$$\frac{17}{12} = \frac{2(8)+1}{2(6)}$$

$$\frac{99}{70} = \frac{2(49)+1}{2(35)}$$

$$\frac{577}{408} = \frac{2(288)+1}{2(204)}$$

If S_N and T_N denote, respectively, the N th square and triangular numbers, then:

$$T_1 = 1 \text{ and } S_1 = 1$$
$$T_8 = 36 \text{ and } S_6 = 36$$
$$T_{49} = 1225 \text{ and } S_{35} = 1225$$
$$T_{288} = 41616 \text{ and } S_{204} = 41616$$

The even terms give squangular numbers!

Challenge. Prove that the ladder gives *all* squangular numbers. (We see that there are infinitely many squangular numbers.)

4. The ladder gives squangular numbers another way: Multiply together the entries in the top and bottom rows:

$$1 \times 1 = 1 \text{ and } S_1 = 1 \text{ is squangular.}$$
$$3 \times 2 = 6 \text{ and } S_6 = 36 \text{ is squangular.}$$
$$7 \times 5 = 35 \text{ and } S_{35} = 1225 \text{ is squangular.}$$
$$17 \times 12 = 204 \text{ and } S_{204} = 41616 \text{ is squangular. And so on.}$$

Challenge. Establish this too!

Comment. Swiss mathematician Leonhard Euler (1707–1783) went a step further and proved that the N th squangular number is given by the formula:

$$\left(\frac{(3+\sqrt{8})^N - (3-\sqrt{8})^N}{2\sqrt{8}} \right)^2$$

Research Corner

a)   Discover other patterns in Theon's Ladder.

b)   Start the ladder with a fraction different from $1/1$. What can you discover?

c)   Create a ladder that approximates $\sqrt{3}$ or $\sqrt{5}$ or $\sqrt{6}$. Any patterns? Are there analogs to the squangular numbers? (What about a ladder for $\sqrt{4}$?)

d)   Is there a number that is simultaneously triangular, square, and pentagonal?

COMMENTARY, SOLUTIONS and THOUGHTS

Indeed, everyone knows that the square root of two is irrational and school children are often given the impression that it is obvious. This is far from the case! Proving that the diagonal of a square is incommensurate with its side length takes some doing.

Greek scholars of the time of Pythagoras (ca. 500 B.C.E.) developed a geometric argument that most likely proceeded in a manner similar to the following ([BOGOMOLNY]):

> If the side of a square and its diagonal were commensurate, then there would be a unit of measure so that each has integral length. Half the square then gives a right isosceles triangle with integer sides. Swing one leg of the triangle along a circular arc as shown (or equivalently, in the modern world, fold a paper cut-out of the triangle along the dotted line) to produce a smaller right isosceles triangle again with integral side lengths. (Make use of similar triangles to see that its side-lengths are indeed integral, or observe what happens if one is folding paper.) We can repeat this process on the smaller and smaller right isosceles triangles that we produce until we obtain a right isosceles triangle with integer side-lengths smaller than the unit of measure! This absurdity shows that our starting assumption of commensurability was wrong.

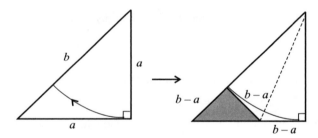

Theodorus of Cyrene (ca. 465-398 B.C.E.) went on and developed additional geometric arguments to establish the irrationality of $\sqrt{n}$ for $n = 3, 5, 6, 7, 8, 10, 11, 12, 13, 14, 15,$ and 17. (See [TANTON1].)

Today many choose to establish the irrationality of $\sqrt{2}$ using a parity argument.

> Suppose $\sqrt{2} = \frac{a}{b}$ with a and b positive integers sharing no common factors. (That is, assume we have written $\sqrt{2}$ as a fraction in reduced form.) Then we must have $2b^2 = a^2$ establishing that a^2, and hence a, is even. Then $2b^2$ is a square of an even number and so is a multiple of four, yielding that b^2, and hence b, is also even. This contradictions the opening assumption.

It is not immediately clear how to generalize this to establish the irrationality of $\sqrt{3}$ and $\sqrt{6}$, yet alone other types of roots ($\sqrt[17]{92}$, for instance).

An effective means of establishing the irrationality of such numbers is to make use of the Fundamental Theorem of Arithmetic:

> *Two equal numbers must have identical prime factorizations (up to the order of the primes).*

It is impossible, for instance, for $7 \times 7 \times 7 \times 11 \times 11 \times 13 \times 41 \times 41 \times 199$ to equal $37 \times 37 \times 37 \times 53 \times 53 \times 53 \times 53 \times 101 \times 101$. (The fundamental theorem was proved by Euclid, ca. 300 B.C.E., and is discussed in [TANTON2].) The theorem leads to the following useful result:

Lemma. *Suppose a and b are two positive integers sharing no common prime factors. If $\frac{a}{b}$ is an integer, then b is 1.*

Proof. If $\frac{a}{b} = n$ for some integer n, then a and bn are two equal numbers possessing different primes in their factorizations, unless $b = 1$ (and $n = a$).   □

We can now classify which roots are rational:

Theorem. *$\sqrt[n]{M}$ is rational if and only if M is the nth power of an integer.*

Proof. If $\sqrt[n]{M} = \frac{a}{b}$ for some fraction in reduced form, then $\frac{a^n}{b^n}$ is an integer, forcing b^n, and hence b, to be 1.   □

Thus $\sqrt{3}$, $\sqrt{6}$, and $\sqrt[17]{92}$ are indeed irrational.

The sum of two irrationals can be rational, for example $\sqrt{2} + (5 - \sqrt{2})$, and a sum of two irrationals can be irrational: as $\sqrt{2} + \sqrt{3}$. (If $\sqrt{2} + \sqrt{3} = \frac{c}{d}$, then $2 + 2\sqrt{6} + 3 = \frac{c^2}{d^2}$ showing that $\sqrt{6} = \frac{c^2 - 5d^2}{2d^2}$ is rational. This is not so.)

Challenge. Prove, using elementary techniques, that $\sqrt{2} + \sqrt{3} + \sqrt{5}$ is irrational. Prove that $\sqrt{2} + \sqrt{3} + \sqrt{5} + \sqrt{7} + \sqrt{11}$ is irrational.

Similarly, multiplying two irrational numbers can lead to both rational and irrational results. Raising irrational numbers to irrational powers, however, is curious and tricky to analyze. Can an irrational number raised to an irrational power ever yield a rational result? The answer is ... YES!

Consider $\sqrt{2}^{\sqrt{2}}$. If it is rational, then we have an example of what we seek. If $\sqrt{2}^{\sqrt{2}}$ is irrational, then $\left(\sqrt{2}^{\sqrt{2}}\right)^{\sqrt{2}}$ is an example because it is an irrational number raised to an irrational power and $\left(\sqrt{2}^{\sqrt{2}}\right)^{\sqrt{2}} = \sqrt{2}^{2} = 2$, a rational! Either way we have an example. However, we don't know if it is $\sqrt{2}^{\sqrt{2}}$ or $\left(\sqrt{2}^{\sqrt{2}}\right)^{\sqrt{2}}$!

Challenge. Is there an example of an irrational number raised to an irrational power that is irrational?

Comment. The *Gelfond-Schneider Theorem*, easy to find on the internet, has something to say about this.

Theon's Ladder

Theon's Ladder ([FLANNERY], [YOUNG]]) is a wonderfully rich and accessible object for mathematical investigation. Students of all ages have explored the many properties of the ladder with me ([TANTON3]) and I am forever astounded by the magic of the sequence and the directions one can go with it. Young students are particularly adept at discovering new and interesting patterns.

Here I outline the key features of the ladder as suggested in the newsletter. The mathematics behind the scenes is Pell's equation, which, in and of itself, offers many rich avenues for deep exploration.

The first thing to note is that Theon's algorithm

$$\frac{a}{b} \rightarrow \frac{a+2b}{a+b}$$

produces from a fraction in reduced form another fraction in reduced form.

If $a + 2b$ and $a + b$ share a common factor d, then so do $(a + 2b) - (a + b) = b$ and $2(a + b) - (a + 2b) = a$. Thus $\gcd(a, b) = 1$ implies $d = 1$.

Theon's Ladder starts with the fraction $\frac{1}{1}$ and produces

a	1	3	7	17	41	99	239	577	1393	...
b	1	2	5	12	29	70	169	408	985	...

Let a_n and b_n denote the n th term of each row. We have

$$a_{n+1} = a_n + 2b_n$$
$$b_{n+1} = a_n + b_n$$

with $a_1 = b_1 = 1$. The terms of $\{a_n\}$ are thus forever odd, and the terms of $\{b_n\}$ alternate in parity. We also see that both sequences are strictly increasing.

Because

$$a_{n+1}^2 - 2b_{n+1}^2 = (a_n + 2b_n)^2 - 2(a_n + b_n)^2 = -\left(a_n^2 - 2b_n^2\right)$$

from $a_1^2 - 2b_1^2 = -1$ we obtain

$$a_n^2 - 2b_n^2 = (-1)^n.$$

Thus $(\frac{a_n}{b_n})^2 = 2 + \frac{(-1)^n}{b_n^2}$ and, since $b_n \rightarrow \infty$ as $n \rightarrow \infty$, we have

$$\frac{a_n}{b_n} \rightarrow \sqrt{2} \text{ as } n \rightarrow \infty.$$

In the newsletter I claimed that the error in the approximation decreases by a factor of at least four from term to term:

$$\left| \frac{a_{n+1}}{b_{n+1}} - \sqrt{2} \right| = \left| \frac{a_n + 2b_n}{a_n + b_n} - \sqrt{2} \right|$$

$$= (\sqrt{2} - 1) \cdot \frac{1}{\frac{a_n}{b_n} + 1} \cdot \left| \frac{a_n}{b_n} - \sqrt{2} \right|$$

$$\leq (\sqrt{2} - 1) \cdot \frac{1}{1 + 1} \cdot \left| \frac{a_n}{b_n} - \sqrt{2} \right|$$

$$< \frac{1}{4} \cdot \left| \frac{a_n}{b_n} - \sqrt{2} \right|,$$

using the fact that $\frac{a_n}{b_n} \geq 1$.

Connections to Pythagorean Triples and Squangular Numbers

We have

$$a_n^2 - 2b_n^2 = (-1)^n$$

which shows that the odd terms of the ladder satisfy

$$x^2 - 2y^2 = -1$$

(with x guaranteed to be odd), which can be rewritten $(\frac{x-1}{2})^2 + (\frac{x+1}{2})^2 = y^2$. For n odd, this reads $(\frac{a_n-1}{2})^2 + (\frac{a_n+1}{2})^2 = (b_n)^2$ giving a Pythagorean triple with legs differing by one among those odd positions of the sequence.

The even terms of the ladder satisfy

$$x^2 - 2y^2 = 1$$

(with x odd and y even), which can be rewritten $\frac{1}{2} \cdot \frac{x-1}{2} \cdot \frac{x+1}{2} = (\frac{y}{2})^2$. If T_n denotes the n th triangular number ($T_n = 1 + 2 + \cdots + n = \frac{n(n+1)}{2}$) and S_n the n th square number ($S_n = n^2$), then this equation reads $T_{\frac{x-1}{2}} = S_{\frac{y}{2}}$. This shows that, for n even, $\frac{a_n-1}{2}$ and $\frac{b_n}{2}$ correspond to the indices of squangular numbers

$$S_{\frac{b_n}{2}} = T_{\frac{a_n-1}{2}} \,(n \text{ even}).$$

The challenge is to prove that every pair of positive integer solutions to $x^2 - 2y^2 = \pm 1$ appears in Theon's Ladder (that is, that every Pythagorean triple with legs differing by 1 and every squangular number appears in the way indicated). Here goes:

Suppose (x, y) is a pair of positive integers satisfying $x^2 - 2y^2 = \pm 1$. Because x and y cannot share a common factor different from 1 (it would be a factor of ± 1 as well), the fraction $\frac{x}{y}$ is in reduced form. We need to show that $\frac{x}{y}$ appears in Theon's Ladder.

If $y = 1$, then it is easy to see that $x^2 - 2y^2 = \pm 1$ gives $x = 1$ as well and we have the term $\frac{1}{1}$ in the ladder.

If $y \neq 1$, then we have some work to do! Let's explore this case.

If $\frac{x}{y}$ appears in the ladder, what would be the term before it? We can check that Theon's algorithm $\frac{a}{b} \rightarrow \frac{a+2b}{a+b}$ implies that $\frac{2y-x}{x-y}$ would be the quantity, but we need this to be a meaningful fraction in the ladder. That is, we need to check that both the numerator and denominator are positive integers and that this fraction is in reduced form.

From $x^2 = 2y^2 \pm 1 = y^2 + (y^2 \pm 1)$ (with $y > 1$) we see that $x > y$. We also see that $x < 2y$ ($x^2 \leq 2y^2 + 1 < 3y^2 < 4y^2$). It is straightforward to check that $2x - y$ and $x - y$ share no common factors if x and y do not. We're good!

From $x^2 - 2y^2 = \pm 1$ it follows that $(2y - x)^2 - 2(x - y)^2 = \mp 1$.

So what have we so far? If (x, y) is a solution to $x^2 - 2y^2 = \pm 1$, then $(2y - x, x - y)$ is another pair of positive integers satisfying the analogous equation and it is a solution with second term smaller than y. Also, if $\frac{x}{y}$ is Theon's ladder, then $\frac{2y-x}{x-y}$ would be the reduced fraction just before it.

We're almost done.

We've established that from one solution $\frac{x}{y}$ we can obtain another solution with smaller denominator. Repeating we can obtain yet another solution with even smaller denominator. Continuing to repeat, we eventually obtain a solution with denominator 1, which we have shown is the term $\frac{1}{1}$ in Theon's Ladder. As all the solutions we are creating come from applying the reverse of Theon's algorithm to $\frac{x}{y}$ to arrive at $\frac{1}{1}$, we have that $\frac{x}{y}$ arises from applying

Theon's algorithm forward starting from $\frac{1}{1}$. Thus $\frac{x}{y}$ appears in the ladder, just as we hoped to show!

Going Further

From

$$a_{n+1} = a_n + 2b_n$$
$$b_{n+1} = a_n + b_n$$

we get

$$a_{n+1} = a_n + b_n + b_n = b_{n+1} + b_n$$

and

$$b_{n+1} = a_n + b_n = b_n + b_{n-1} + b_n = 2b_n + b_{n-1}$$

so

$$b_{n+1} = 2b_n + b_{n-1}$$

with $b_1 = 1$ and $b_2 = 2$. It is convenient to set $b_0 = 0$.

What sequences satisfy this recursion relation?

It is a standard technique to examine whether or not a geometric sequence 1, x, x^2, x^3, x^4, ... could satisfy a given recursion relation. In our case we are looking for a value of x that satisfies

$$x^{n+2} = 2x^{n+1} + x^n,$$

that is, one that satisfies $x^2 = 2x + 1$. There are two solutions: $x_1 = 1 + \sqrt{2}$ and $x_2 = 1 - \sqrt{2}$.

Thus we have two geometric sequences that fit the recurrence relation:

$$1, \ x_1, \ x_1^2, \ x_1^3, \ x_1^4, \ \ldots \text{ and } 1, \ x_2, \ x_2^2, \ x_2^3, \ x_2^4, \ \ldots$$

It is straightforward to check that a linear combination of the two sequences satisfies the same relation, so

the sequence $\alpha + \beta, \alpha x_1 + \beta x_2, \alpha x_1^2 + \beta x_2^2, \alpha x_1^3 + \beta x_2^3, \alpha x_1^4 + \beta x_2^2, \ldots$

satisfies $\alpha x_1^{n+2} + \beta x_2^{n+2} = 2 \left(\alpha x_1^{n+1} + \beta x_2^{n+1} \right) + \left(\alpha x_1^n + \beta x_2^n \right).$

Choose values for α and β that give a sequence with initial values $b_0 = 0$ and $b_1 = 1$: $\alpha = \frac{1}{2\sqrt{2}}$ and $\beta = -\frac{1}{2\sqrt{2}}$ do the trick.

Now ... if two sequences satisfy the same recursion relation and have the same initial conditions, as we have here, they are the same sequence. We have thus established

$$b_n = \frac{(1 + \sqrt{2})^n - (1 - \sqrt{2})^n}{2\sqrt{2}}.$$

Algebra gives:

$$a_n = b_n + b_{n-1} = \frac{(1 + \sqrt{2})^n + (1 - \sqrt{2})^n}{2}.$$

Challenge. Find a formula for $\frac{a_n}{b_n}$ and use it to show that $\frac{a_n}{b_n} \to \sqrt{2}$ as $n \to \infty$.

More algebra shows that $\frac{b_{2n}}{2}$ and $a_n b_n$ both equal

$$\frac{(3 + \sqrt{8})^n - (3 - \sqrt{8})^n}{2\sqrt{8}}$$

so $2a_n b_n = b_{2n}$. This gives Euler's result:

From $S_{\frac{b_{2n}}{2}} = T_{\frac{a_{2n}-1}{2}}$, the n th squangular number is

$$\left(\frac{b_{2n}}{2}\right)^2 = \left(\frac{(3 + \sqrt{8})^n - (3 - \sqrt{8})^n}{2\sqrt{8}}\right)^2.$$

Final Thoughts. Starting Theon's Ladder with the fraction $\frac{5+12}{13} = \frac{17}{13}$ generates among its odd terms all Pythagorean triples with legs that differ by seven. What is hiding among the even terms?

Is it possible to extend Theon's Ladder infinitely far to the left? Are there geometric interpretations for the negative terms?

I do not know of a number that is simultaneously triangular, square, and pentagonal, nor have any of my students found one. Might you find one?

References

[BOGOMOLNY] Bogomolny, A., Cut the Knot! The Square Root of Two is Irrational. URL: www.cut-the-knot.org/proofs/sq_root.shtml.

[FLANNERY] Flannery, D., *The Square Root of 2*, Praxis Publishing Ltd, New York, NY, 2006.

[TANTON1] Tanton, J., *Encyclopedia of Mathematics,* Facts on File, New York, NY, 2005.

[TANTON2] Tanton, J. *THINKING MATHEMATICS! Volume 1: Arithmetic = Gateway to All,* www.lulu.com, 2009.

[TANTON3] Tanton, J., Pit your wits against young minds! *Mathematical Intelligencer*, **29** no. 3, (2007), 55–59.

[YOUNG] Young, R., *Excursions in Calculus: An Interplay of the Continuous and the Discrete,* The Mathematical Association of America, Washington D.C., 1992.

24

Tilings and Theorems

TIDBIT: Clever Tessellations

1. Here's a picture of a portion of bathroom floor tiled with three differently shaped tiles: a big square, a small square, and a parallelogram.

Place a dot at the center of each white square. (Really. Please do it!) Do you see from this we get an easy proof of the following remarkable result from geometry?

   If squares are drawn on each edge of a parallelogram, their centers form a perfect square.

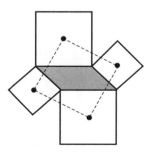

2. Pythagoras's theorem states that if squares are drawn on the sides of a right triangle, then the sum of areas of the small squares equals the area of the large square.

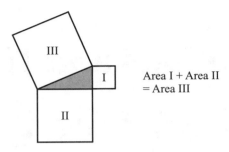

Area I + Area II
= Area III

Can you see that this tiling of a bathroom floor provides a visual proof of the Pythagorean theorem?

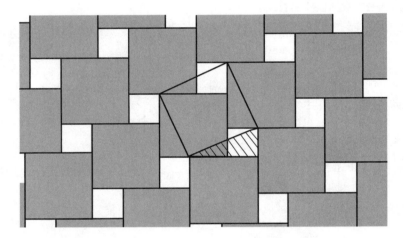

(The white tile is square I, the shaded tile is square II, and the tilted figure is square III. The tilted square is composed of five pieces—two white and three shaded—that together make squares I and II!)

3. The French Emperor Napoleon was fond of geometry. He discovered and proved the following remarkable result:

    On each side of a triangle draw an equilateral triangle. Then the centers of the equilateral triangles form a perfect equilateral triangle.

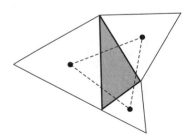

The result can be proved by tiling the floor with congruent copies of the same scalene triangle. The centers of the equilateral triangular spaces form a triangular lattice!

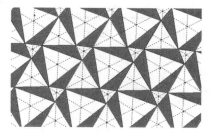

Cutting out large copies of shapes in poster board and tiling part of the floor to look for hidden patterns makes for a great classroom activity. Here students from the St.Mark's School tile the floor with copies of a single scalene triangle and place Hershey Kisses in the equilateral triangle spaces to make the regular pattern apparent.

PUZZLER: Squaring Off

In this picture each line connects a vertex of the square to the midpoint of an opposite side. If the area of the original square is one unit, what is the area of the interior square?

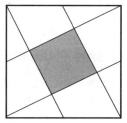

Lines are drawn from points one third along the sides of the square as shown. If the area of the original square is one unit, what is the area of the inner squares?

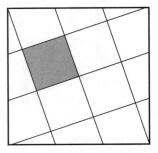

In general, if each side of a square is divided into N parts and lines are drawn across the figure in the general manner suggested, what is the area of the inner squares?

Research Corner. Not many people realize that Pythagoras's theorem also works for equilateral triangles.

   Draw equilateral triangles on the three sides of a right triangle. The area of the large triangle equals the sum of the areas of the two smaller triangles

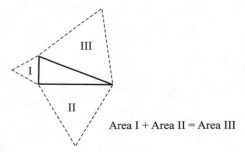

Area I + Area II = Area III

Is there a way to combine the second and third tiling schemes of the opening essay to produce a visual proof of this result? Can one develop visual proofs for other shapes drawn on sides of right triangles?

COMMENTARY, SOLUTIONS and THOUGHTS

There is a tremendous amount of mathematics that can be deduced from inspection of clever tiling patterns. Here are three results in addition to the ones presented in the newsletter. (See [RIGBY] and [NELSON] for more such proofs.)

Theorem 1. *The area of a quadrilateral is half the area of the parallelogram formed by its diagonals.*

Proof. The result is easily seen true for a convex quadrilateral if we enclose it in the parallelogram formed by its diagonals:

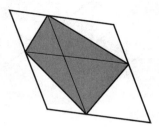

For a concave quadrilateral, tile(!) and inspect the diagonals:

□

Theorem 2. *For any four positive real numbers* $a, b, u,$ *and* v :

$$au + bv \leq \sqrt{a^2 + b^2} \cdot \sqrt{u^2 + v^2}.$$

Proof. Draw two rectangles with dimensions as shown

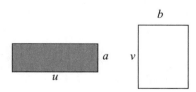

and use them in a tessellation.

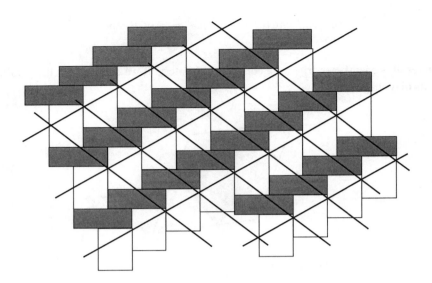

Connecting the lower corners of the shaded rectangles produces a grid of parallelograms in which each parallelogram has area equal to the sum of areas of two rectangles

$$area = au + bv.$$

If we draw two adjacent parallelograms, we see that the side-lengths of a single parallelogram are $\sqrt{a^2 + b^2}$ and $\sqrt{u^2 + v^2}$.

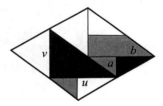

Thus the area of a single parallelogram is also given by

$$area = \sqrt{a^2 + b^2} \cdot \sqrt{u^2 + v^2} \cdot \sin\theta,$$

where θ is one of the interior angles of the parallelogram.

Consequently

$$au + bv = \sqrt{a^2 + b^2} \cdot \sqrt{u^2 + v^2} \cdot \sin\theta.$$

Since $\sin\theta \leq 1$ the result follows.   □

Question. In vector calculus the quantities $\sqrt{a^2 + b^2}$ and $\sqrt{u^2 + v^2}$ represent the magnitudes of the vectors $\bar{v} = <a, b>$ and $\bar{u} = <u, v>$, and $\bar{v} \cdot \bar{u} = au + bv$ is their dot product. We know $\bar{v} \cdot \bar{u} = ||\bar{v}||\,||\bar{u}|| \cos\theta$. In the proof we seem to have established $\bar{v} \cdot \bar{u} = ||\bar{v}||\,||\bar{u}|| \sin\theta$. Can you explain the discrepancy?

This third result is charming:

Theorem 3. *Inscribe within a circle a square and a rectangle with one side twice the length of the other. Then the rectangle has four-fifths the area of the square.*

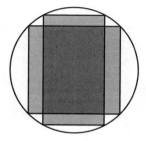

Proof. A square tiling reveals all:

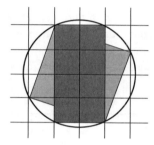

□

Challenge. A square and a rectangle with one side three times the length of the area are both inscribed in a circle. What can you say about their areas?

Develop a whole host of results of this type.

Squaring Off

To answer the puzzler of the newsletter, place two copies of the subdivided squares next to each other to see copies of the central square repeated.

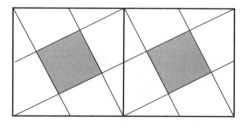

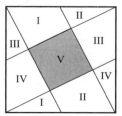

It is now clear that lines connecting midpoints create a central shaded square one-fifth the area of the original square.

For lines connecting one-third points along the side, we see that a small square has area one-tenth the area of the original square.

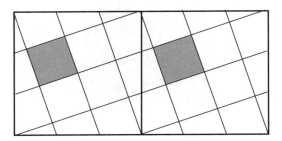

In general, subdividing the sides of a square each into n equal parts and connecting lines across the figure in an analogous manner creates subsquares of fraction $\frac{1}{n^2+1}$ the area.

Challenge. What is the area of the shaded square shown in terms of x?

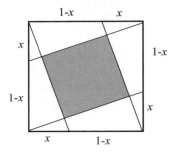

These ideas extend to triangles as well. For example, suppose three lines are drawn inside a triangle each connecting a vertex to a point one third of the way along the opposite side.

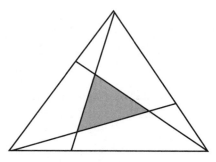

The area of the shaded triangle has area one seventh the area of the original triangle. A triangular tessellation shows this is so.

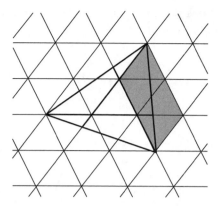

Each side of the large triangle is the diagonal of a parallelogram composed of four triangles. We see that the area of the large triangle matches the area of seven small triangles.

Challenge. Check that this result holds for all triangles by drawing a triangle that, unlike the triangle above, looks very scalene and draw the tessellation that accompanies it.

Challenge. Suppose, instead, lines are drawn connecting each vertex of a triangle to a point one quarter of the way along the opposite side. Show that the area of the central triangle created is now four-thirteenths the area of the original triangle.

Generalize!

Tough Challenge. Consider an equilateral triangle with sides of length 1. Is there a formula, in terms of x, y and z for the area of the central triangle?

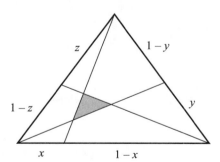

(Are there instances for which this area should be regarded as negative?)

Comment. *Ceva's Theorem* in geometry gives that the area of the triangle is zero if $xyz = (1-x)(1-y)(1-z)$.

On Pythagoras's Theorem

Recall that if we scale a picture of a figure by a factor k, then its area changes by a factor k^2. This has a remarkable consequence.

Suppose we take a figure with a side of length one unit and area A and scale it with factors a, b, and c. This produces figures of areas $a^2 A$, $b^2 A$ and $c^2 A$.

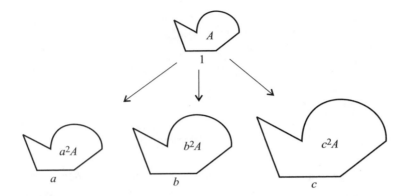

Placing these figures along the edges of a right triangle with sides of lengths a, b, and c creates a Pythagorean-like diagram with area I $= a^2 A$, area II $= b^2 A$ and area III $= c^2 A$.

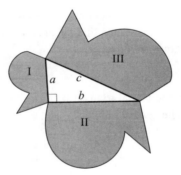

Because $a^2 + b^2 = c^2$, we have that

$$\text{area I} + \text{area II} = \text{area III}$$

which shows that

Pythagoras's Theorem holds for any set of similar figures drawn on the sides of a right triangle.

This is astounding.

Comment. The general observation also applies to equilateral figures drawn on the sides of a right triangle. See [ALSINA and NELSEN] for a purely visual proof of the theorem for equilateral triangles. Care to create analogous visual proofs for other equilateral shapes?

Challenge. Equilateral triangles are drawn on the sides of a right triangle. Let P be the point along the hypotenuse of the right triangle so that the large equilateral triangle is divided into two pieces with areas I and II as shown. This point P has a special property. What?

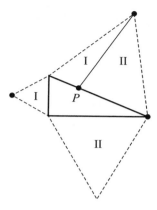

What if squares are drawn on the sides of the right triangle instead? Show that the same point P on the hypotenuse marks a location that divides square III into two rectangles of areas equal to the areas of square I and square II. (See [TANTON, Appendix V].)

Final Challenge. Napoleon's Theorem holds true if, instead of drawing equilateral triangles on the outside of each edge of a given triangle, we draw equilateral triangles on the <u>inside</u> of each edge. This produces a diagram that overlaps on itself, but the centers of those equilateral triangles again form an equilateral triangle. Is there any way to demonstrate this version of the result with tiling?

References

[ALSINA and NELSEN] Alsina, C. and Nelsen, R., Proof Without Words: The Pythagorean Theorem with Equilateral Triangles, *College Mathematics Journal*, **43**, no. 3, (2012), 226.

[NELSEN] Nelsen, R., *Proofs Without Words II: More exercises in Visual Thinking*, The Mathematical Association of America, Washington D.C., 2000.

[RIGBY] Rigby, J., Fun with tessellations in *The Lighter Side of Mathematics: Proceedings of the Eugène Strens Memorial Conference on Recreational Mathematics and its History*. R. Guy and R. Woodrow, editors, Mathematical Association of America, Washington D.C., 1996.

[TANTON] Tanton, J., *Geometry, Volume II*, www.lulu.com, 2009.

25

The Tower of Hanoi

PUZZLER: A Classic

The Tower of Hanoi puzzle consists of three poles on which a collection of differently sized discs sit on one pole in order of size, largest on the bottom. The challenge is to transfer all the discs to a different pole in such a way that only one disc is moved at a time and no disc throughout the process ever sits on top of a disc of smaller size.

The two-disc version of the puzzle is easy to solve. It takes three moves.

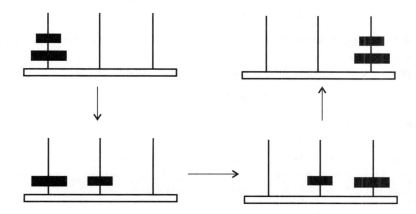

a) Solve the three-disc version of the puzzle. How many moves does it take?

An eight-disc version of this puzzle was patented and sold as a toy by Edouard Lucas in 1883. He claimed it was based on a "tower of Brahma" in which priests were given the task of transferring 64 discs from one pole to another, one disc per day. Legend has it that the world will end on the day the task is completed.

b) If priests started the 64-disc challenge on day one, on which day will the world end?

c) If the discs are numbered 1 through 64 with 1 being the smallest disc and 64 the largest, which disc will the priests move on the 10,000th day? (Assume that they follow the most efficient means of solution.)

The Tower of Hanoi. There are a number of mathematical surprises hidden in the analysis of Lucas' puzzle. Consider an n-disc version of the puzzle, with discs numbered 1 for the smallest to n for the largest. To solve the one-disc puzzle, one just has to move disc number 1:

$$1$$

To solve the two-disc puzzle, one moves disc 1, then 2, and then disc 1 again:

$$1, 2, 1$$

Here is the order of discs one must move to solve the three-, and four-disc puzzles:

$$1, 2, 1, 3, 1, 2, 1$$

$$1, 2, 1, 3, 1, 2, 1, 4, 1, 2, 1, 3, 1, 2, 1$$

In general, to solve the n-disc puzzle, one must first move the top $(n - 1)$ discs to a second pole, move disc n and then move the $(n - 1)$ discs on top of the large disc. This means following the solution to the $(n - 1)$ -disc puzzle twice.

$$\text{new sequence of moves} = \text{previous sequence} + n + \text{previous sequence}$$

The number of moves required to solve the n-disc puzzle is double the previous count of moves plus one. This generates the numbers that are one less than a power of two:

$$1, 3, 7, 15, 31, 63, 127, 255, 511, 1023, \ldots$$

[CHECK THIS!: If n is one less than a power of two then $2n + 1$ is also one less than a power of two.]

Thus, according to the legend of the Brahmas, the world will end on day

$$2^{64} - 1 = 18, 446, 744, 073, 709, 551, 615,$$

that is, in about 5.05×10^{16} years.

Question. Suppose all the discs are first stacked on the leftmost pole and we wish to end with them stacked on the rightmost pole. To which pole – middle or rightmost – should we first move disc 1 to accomplish this? Does the answer depend on the number of discs?

BONUS PUZZLER: Another Connection to Powers of Two

Consider the sequence of disc moves for the four-disc solution:

$$1, 2, 1, 3, 1, 2, 1, 4, 1, 2, 1, 3, 1, 2, 1$$

(This is also the beginning of the five-disc sequence, and the beginning of the 17-disc solution, and so on. For this reason we can say that there is a well-defined <u>infinite</u> tower of Hanoi sequence of moves.)

Subtract one:

$$0, 1, 0, 2, 0, 1, 0, 3, 0, 1, 0, 2, 0, 1, 0$$

Now:

2^0 is the largest power of two that divides 1

2^1 is the largest power of two that divides 2

2^0 is the largest power of two that divides 3

2^2 is the largest power of two that divides 4

2^0 is the largest power of two that divides 5

2^1 is the largest power of two that divides 6

2^0 is the largest power of two that divides 7

2^3 is the largest power of two that divides 8

and so on. Since $16 = 2^4$ is the largest power of two that divides 10,000 this means that the priests will be moving disc $4 + 1 = 5$ of their 64-disc puzzle on day 10,000.

Challenge. Why this connection with the powers of two?

Sierpinski-type Pictures. There are nine possible ways to place three discs on the three poles. If we display all nine configurations and connect two with a line if one can be reached from the other by the move of a single disc we obtain the following display:

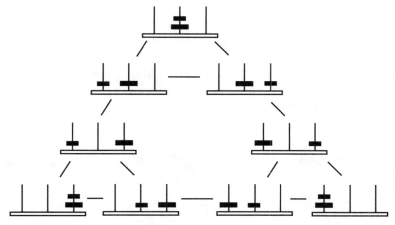

There are 27 possible ways to place four discs on three poles. If we had the patience to display all 27 configurations connecting any two that can be obtained from one another by the move of a single disc we'd obtain the following display. (Do it!)

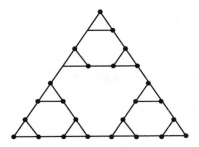

With five discs we obtain:

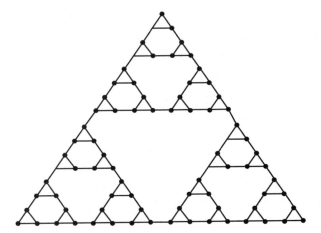

These pictures make sense: Each component of the diagram for n discs corresponds to the placement of the largest disc. As there are three possible locations, there are three components. With the largest disc in place, each component must be a replica of the previous $(n-1)$-disc picture. Also, as there is only one way to move the largest disc from one specific location to another, two components are linked by only one edge.

Research Corner. As we have seen, the minimum number of moves required to move n discs from one pole to another among three poles is $2^n - 1$. No one knows a formula for the minimum number of moves required to move n discs from one pole to another among four poles. One disc can be moved in 1 move, two discs in 3 moves, three in 5 moves, four in 9 moves, five in 13 moves, six in 17, seven in 25, eight in 33. Any insights or thoughts?

COMMENTARY, SOLUTIONS and THOUGHTS

For a detailed history of the Tower of Hanoi puzzle see [HINZ].

We saw in the newsletter that for n discs (numbered from smallest to largest as 1 through n) on three pegs, the sequence of moves that lead to a solution is built from the solution sequence on just $n-1$ discs:

$$\begin{matrix} \text{new sequence} \\ \text{of moves} \end{matrix} = \begin{matrix} \text{previous} \\ \text{sequence} \end{matrix} + n + \begin{matrix} \text{previous} \\ \text{sequence} \end{matrix}$$

Starting with 1 (for the one-disc puzzle) this relation generates

$$1 \to 121 \to 1213121 \to 121312141213121 \to \cdots$$

each the start of a single, well-defined infinite tower of Hanoi sequence:

$$1213121412131215121312141213121 6\cdots$$

From the generating relation, every odd term of this sequence is 1. A little thought shows that the even terms satisfy the same generating relation, except starting with 2.

even terms of (n+1)th sequence	=	even terms of n-th sequence	+	n+1	+	even terms of n-th sequence

$$\_ \to \_2\_ \to \_2\_3\_2\_ \to \_2\_3\_2\_4\_2\_3\_2\_ \to \cdots$$

Thus the sequence of even terms within the infinite tower of Hanoi sequence is the same as the entire original sequence but with all entries one larger.

If $t(n)$ denotes the nth entry of the infinite tower of Hanoi sequence, we have thus shown

$$t(2n + 1) = 1 \text{ and } t(2n) = t(n) + 1.$$

It follows that if we write $n = 2^r m$ with m odd, then $t(n) = t(m) + r = r + 1$. Thus

> The nth entry of the infinite tower of Hanoi sequence is one more than the largest power of two that divides n.

For a wonderful discussion on the graphs associated with tower of Hanoi puzzles, some known facts and many open questions see [ARETT and DORÉE].

Final Comment. To move a stack of n discs from the left pole to the right pole, start by moving the small disc to the right pole if n is odd and to the middle pole if n is even.

References

[ARETT and DORÉE] Arett, D., and Dorée, S., Coloring and counting on the tower of Hanoi graphs, *Mathematics Magazine,* **83** (2010), 200–209.

[HINZ] Hinz, A. M., The tower of Hanoi, *Enseignement Mathematique (2)*, **35** (1989), 289–321.

26

Weird Multiplication

PUZZLER: Weird Multiplication

Here's an unusual means for performing long multiplication. To compute 22×13, for example, draw two sets of vertical lines, the left set containing two lines and the right set two lines (for the digits in 22) and two sets of horizontal lines, the upper set containing one line and the lower set three (for the digits in 13).

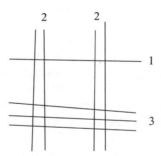

There are four sets of intersection points. Count the number of intersections in each and add the results diagonally as shown:

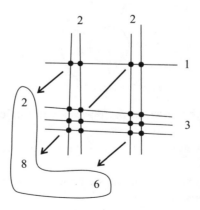

The answer 286 appears.

There is one caveat as illustrated by the computation 246 × 32:

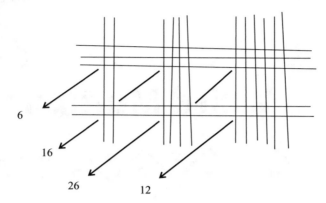

Although the answer 6 thousands, 16 hundreds, 26 tens, and 12 ones is absolutely correct, we need to carry digits and translate this into 7,872.

a)   Compute 131 × 122 by this method.

b)   Compute 54 × 1332 by this method.

c)   How best should one compute 102 × 30054 by this method?

d)   Why does the method work?

A Tiny Tidbit

Putting on one's shoes and then one's socks does not produce the same result as first putting on one's socks and then one's shoes. These two operations are not commutative.

In mathematics, multiplication is said to be commutative, meaning that $a \times b$ produces the same result as $b \times a$ for all numbers a and b. Is this obviously true? It might not be, given the following graphical model for multiplication:

To compute 2×3, for instance, first draw two concentric circles to model the first number in the problem, 2, and three radii to model the second number, 3.

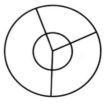

The number of separate pieces one sees in the diagram—in this case 6—is the answer to the multiplication problem.

Is it at all obvious that drawing three concentric circles and two radii will produce the same number of pieces?

Suppose we draw 77 circles and 43 radii. Can we be sure that drawing 43 circles and 77 radii yields an equivalent number of pieces?

Is the commutativity of multiplication a belief or a fact?

Another Tidbit

Here's another way to compute a multiplication problem; 341 × 752, for instance.

Write each number on a strip of paper, reversing one of the numbers. Start with the reversed number on the top strip to the right of the second number.

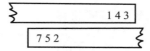

Slide the top number to the left until a first set of digits align. Multiply them and write their product underneath.

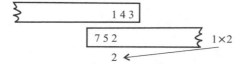

Slide one more place to the left and multiply the digits that are aligned. Write their sum underneath.

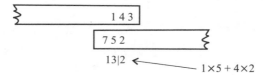

Continue:

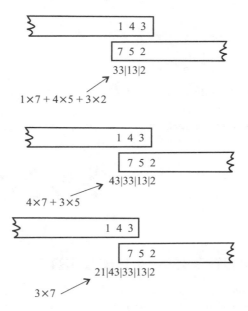

The answer to 341 × 752 is:

$$21 \mid 43 \mid 33 \mid 13 \mid 2$$

namely, 21 ten-thousands, 43 thousands, 33 hundreds, 13 tens and 2 ones. After carrying digits this becomes 256, 432.

Comment. Some people may prefer to carry digits as they conduct this procedure and write something like:

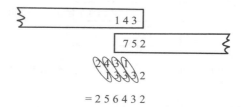

$$= 2\,5\,6\,4\,3\,2$$

Another Puzzler

Vedic mathematics taught in India (and established in 1911 by Jagadguru Swami Bharati Krishna Tirthaji Maharaj) has students compute the multiplication of two three-digit numbers as follows:

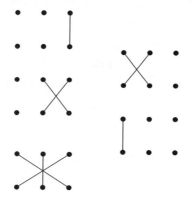

What do you think the sequence of diagrams means?

Research Corner. Invent your own method of conducting long multiplication.

COMMENTARY, SOLUTIONS and THOUGHTS

I have found the content of this newsletter to be a big hit with teachers of middle-school students and, in general, a delight for students of all ages. Along with "finger multiplication" and "Russian multiplication" presented below, these techniques truly demonstrate the joy to be had by engaging in intellectual play.

The multiplication methods outlined invite us to reexamine the long multiplication algorithm we teach our students. Consider, for example, the computation 83×27. Students might be taught to write something like

$$
\begin{array}{r}
8\ 3 \\
\times\ 2\ 7 \\
\hline
2\ 1 \\
56\ 0 \\
6\ 0 \\
160\ 0 \\
\hline
22\ 4\ 1
\end{array}
$$

Mathematicians recognize this as an exercise in expanding brackets:

$$83 \times 27 = (80 + 3)(20 + 7) = 1600 + 60 + 560 + 21.$$

Students might be taught to associate an area model with multiplication where expanding brackets is equivalent to dividing a rectangle into pieces.

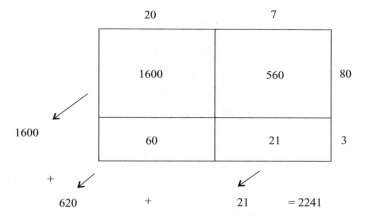

Adding the numbers in the cells according to the diagonals on which they lie groups the 100s, the 10s and the units.

Comment. When I teach long multiplication to students, I use the area model. I feel that the level of understanding that comes from it far outweighs the speed that comes from the standard algorithm: students remain aware that the "8" in the above problem means eighty and there is no mystery about placement of zeros and carrying digits.

Aside. In the 1500s in England students were taught to compute long multiplication using the following *galley method* (also known as the *lattice method* or the *Elizabethan method*). (See [READERS]):

> *To multiply 218 and 43 draw a 2 × 3 grid of squares. Write the digits of the first number along the top of the grid and the digits of the second number along the right side. Divide each cell of the grid diagonally and write in the product of the column digit and row digit of that cell, separating the tens from the units across the diagonal of that cell. (If the product is a one digit answer, place a 0 in the tens place.)*

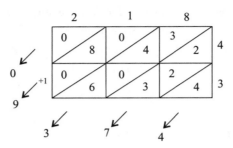

Add the entries in each diagonal, carrying tens digits over to the next diagonal if necessary, to see the final answer. In the example, $218 \times 43 = 9374$.

The diagonals group together like powers of ten. (The lack of zeros can make this process mysterious to young students.)

It is now easy to see that the lines method is the area model of multiplication in disguise, since the count of intersection points matches the multiplication of the single digit numbers.

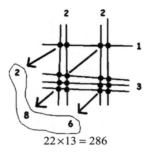

$$22 \times 13 = 286$$

The strip method is a geometric tracking of the individual multiplications that occur in a long multiplication problem (with place-value keeping track of the powers of ten for us). And the Vedic technique ([BATHIA])) is a memorization device for the standard algorithm.

Circle/Radius Math

Drawing a common radii for b circles yields ab pieces because one radius produces b pieces and each radius inserted thereafter adds an additional b segments.

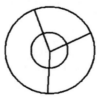

Thus, circle/radius math is ordinary multiplication in disguise. (Is it? What happens if one or both of a or b is zero?) As ordinary multiplication is commutative, this multiplication is too. Without thinking through the arithmetic, this can be a surprise.

Challenge. Describe a geometric transformation that converts a diagram of *a* common radii for *b* circles into one of *b* common radii for *a* circles.

PEDAGOGICAL COMMENT: The Role of the Area Model for Multiplication

Mathematicians recognize that the area model for multiplication has its theoretical limitations—what is the area of a $-\sqrt{2}$ by $3 + 2i$ rectangle, for example?—but it does justify the axioms we work with when defining a ring. Pedagogically, it seems to be an appropriate and intuitively accurate place for beginning understanding and investigation.

For example, if we allow for negative numbers and consider rectangles with negative side-lengths (young students delight in breaking the rules in this way), then we can help justify many confusing features of the arithmetic we choose to believe. My favorite is why negative times negative should be positive.

For counting numbers, it seems natural to say that 7×3 represents seven groups of three and so equals 21. In the same way, $7 \times (-3)$ represents seven groups of negative three and equals -21. It is difficult to interpret $(-7) \times 3$ in this way, but if we choose to use commutativity, then we can say that this is the same as three groups of negative seven and so again is -21. Up to this point, I find that students (and teachers!) feel at ease and willing to say that all is fine. The kicker comes in trying to understand $(-7) \times (-3)$.

Let's do this by computing 23×17 multiple ways. Here are three:

	30	7
20	600	140
3	90	21

$$23 \times 37 = 600 + 90 + 140 + 21 = 851$$

	30	7
30	900	210
3	−210	−49

$$23 \times 37 = 900 - 210 + 210 - 49 = 851$$

	40	−3
20	800	−60
3	120	−9

$$23 \times 37 = 800 + 120 - 60 - 9 = 851$$

A fourth, being the same computation, must yield the same answer:

	40	−3
30	1200	−90
−7	−280	??

$$23 \times 37 = 1200 - 280 - 90 + ?? = 851$$

We thus have no choice but to set $(-7) \times (-3)$ equal to $+21$.

Question. How do the axioms of a ring ensure that negative times negative must be positive in $\mathbb{Z}$ Where are those axioms in play in the area model argument?

Two More Multiplication Tidbits . . .

Each of these curiosities appeared in newsletters. They also appear in [TANTON1].

1. Finger Multiplication. Don't memorize your multiplication tables. Let your fingers do the work! If you know 1×1 up to 5×5, then there is an easy way to compute product values in the six-through ten times tables. First say that

> *A closed fist represents five and a finger raised adds one.*

Thus a hand with two fingers raised represents seven and a hand with three fingers raised represents eight.

To multiply two numbers between five and ten . . .

a) *Encode the two numbers, one on each hand, and count ten for each finger raised.*

b) *Count the number of* unraised *fingers on each hand and multiply together the two counts.*

c) *Add the results of steps one and two. This is the desired product.*

For example, seven times eight is represented as two raised fingers on the left hand and three on the right hand. There are five raised fingers in all, yielding 50 for step one. The left hand has three lowered fingers and the right, two. Because $3 \times 2 = 6$, the product is $50 + 6 = 56$.

Similarly, nine times seven is computed as $60 + 1 \times 3 = 63$, and nine times nine as $80 + 1 \times 1 = 81$. We never have to multiply any numbers greater than five!

Fingers and Toes. We can compute larger products using the same method! For example, with fingers and toes, we interpret 17×18 as seven raised fingers and eight raised toes. This time we count each raised digit as twenty (we have twenty digits fingers and toes in all) yielding $17 \times 18 = 20 \times 15 + 3 \times 2 = 306$.

Question. Martians have six fingers on each of two hands. Describe their version of the finger multiplication trick.

2. Russian Multiplication. The following unusual method of multiplication is believed to have originated in Russia. (Though it is based on a method of multiplication described by Ahmes in the Rhind papyrus, ca. 1650 B.C.E. See [JOSEPH].)

a) *Head two columns with the numbers you wish to multiply.*

b) *Progressively halve the numbers in the left column (ignoring remainders) while doubling the figures in the right column. Reduce the left column to one.*

c) *Delete all rows with an even number in the left and add all the numbers that survive in the right. This sum is the desired product.*

$$
\begin{array}{ll}
37 \times & 23 \\
\cancel{18} & \cancel{46} \\
9 & 92 \\
\cancel{4} & \cancel{184} \\
\cancel{2} & \cancel{368} \\
1 & 736 \\
\hline
& \boxed{851}
\end{array}
$$

$$37 \times 23 = 851$$

See also [TANTON2] for details.

Final Thought. This chapter is about weird multiplication methods. But have you noticed that the spelling of the word "weird" is itself weird?

References

[BATHIA] Bathia, D., *Vedic Mathematics Made Easy*, Jaico Publishing House, Mumbai, India, 2005.

[JOSEPH] Joseph, G.G., *The Crest of the Peacock: Non-European Roots of Mathematics*, Princeton University Press, New Jersey, 2000.

[READERS] Reader's Digest, *Oddities: In Words, Pictures and Figures,* Reader's Digest Services, Sydney, Australia, 1975.

[TANTON1] Tanton, J., *THINKING MATHEMATICS! Volume1:Arithmetic = Gateway to All,* www.lulu.com, 2009.

[TANTON2] Tanton, J., A dozen questions about the powers of 2, *Math Horizons*, September 2001, 5–10.

Appendix

Numbers that are the Sum of Two Squares

One of the many joys and pleasures in working with young students on a varied array of topics is that they spur you to think about a wide spectrum of mathematical questions. During our work on the Stern-Brocot tree one semester (newsletter 21) I wondered if it is possible to use the tree to prove that every prime one greater than a multiple of four is a sum of two squares. (I could see a connection between the continued fractions we were exploring and the numerators of certain fractions that appeared in the tree.) These thoughts did not apply to where the students wanted to go and were at a level too advanced for the group, so I didn't share them with the class. But I did explore them.

I didn't end up making use of the tree directly in my thinking, but I came to a point where I could prove the classification result mentioned in newsletter 12.

N is a sum of two squares if and only if each prime that is one less a multiple of four (3, 7, 11, 19, 23, ...) that appears in the prime factorization of N does so an even number of times.

This is a classic result in number theory discussed – and proved – in many texts. (See [HARDY and WRIGHT], for example.)

The following represents a series of steps that establishes the classification, indirectly inspired by students of the St. Mark's Math Institute. Proving that every prime congruent to one modulo four is a sum of two squares is a key component.

Finite Continued Fractions: The Basics

Given a sequence of positive reals $a_0, a_1, a_2, \ldots$ let $[a_0, a_1, \ldots, a_n]$ denote the quantity

$$a_0 + \cfrac{1}{a_1 + \cfrac{1}{a_2 + \cdots + \cfrac{1}{a_n}}}.$$

Set $p_0 = a_0$, $q_0 = 1$ and $p_1 = a_0a_1 + 1$, $q_1 = a_1$ and for $n \geq 1$ define

$$p_{n+1} = p_n a_{n+1} + p_{n-1},$$
$$q_{n+1} = q_n a_{n+1} + q_{n-1}.$$

Lemma 1. $[a_0, a_1, \ldots, a_n] = \frac{p_n}{q_n}$ *for all* $n \geq 0$.

Proof. We see this is true for $n = 0$ and $n = 1$. Suppose it is true for $n = m$. That is, we have:

$$[a_0, a_1, \ldots, a_{m-1}, a_m] = \frac{p_m}{q_m} = \frac{p_{m-1}a_m + p_{m-2}}{q_{m-1}a_m + q_{m-2}}.$$

The values p_{m-1}, p_{m-2}, q_{m-1} and q_{m-2} depend only on the terms $a_0, a_1, \ldots, a_{m-1}$. Thus we can say more generally that

$$[a_0, a_1, \ldots, a_{m-1}, x] = \frac{p_{m-1}x + p_{m-2}}{q_{m-1}x + q_{m-2}}$$

for any positive real x. Then

$$
\begin{aligned}
[a_0, a_1, \ldots, a_{m-1}, a_m, a_{m+1}] &= \left[a_0, a_1, \ldots, a_{m-1}, a_m + \frac{1}{a_{m+1}}\right] \\
&= \frac{p_{m-1}\left(a_m + \frac{1}{a_{m+1}}\right) + p_{m-2}}{q_{m-1}\left(a_m + \frac{1}{a_{m+1}}\right) + q_{m-2}} \\
&= \frac{a_{m+1}(p_{m-1}a_m + p_{m-2}) + p_{m-1}}{a_{m+1}(q_{m-1}a_m + q_{m-2}) + q_{m-1}} \\
&= \frac{a_{m+1}p_m + p_{m-1}}{a_{m+1}q_m + q_{m-1}} \\
&= \frac{p_{m+1}}{q_{m+1}}
\end{aligned}
$$

which completes the induction step of the proof.   □

When $a_0, a_1, a_2, \ldots$ is a sequence of positive <u>integers</u> (except $a_0 = 0$ is allowed) $[a_0, a_1, \ldots, a_n]$ is a fraction, called a (finite) continued fraction, which we have shown equals $\frac{p_n}{q_n}$. The recursion relations show that the sequences $\{p_n\}$ and $\{q_n\}$ are increasing: $a_0 = p_0 < p_1 < p_2 < \cdots$ and $1 = q_0 < q_1 < q_2 < \cdots$.

Corollary 2. *The integers* p_n *and* q_n *are coprime for all* n.

Proof. The recursion relations show, after some algebraic work, that $p_n q_{n-1} - p_{n-1}q_n = -(p_{n-1}q_{n-2} - p_{n-2}q_{n-1})$. From $p_1 q_0 - p_0 q_1 = (a_0 a_1 + 1) \cdot 1 - a_0 a_1 = 1$ it follows that

$$p_n q_{n-1} - p_{n-1}q_n = (-1)^{n-1}$$

for all $n \geq 1$. Thus if p_n and q_n share a common factor d, we have $d \mid (-1)^{n-1}$, so $d = 1$.   □

Thus $\frac{p_n}{q_n}$ is the fraction $[a_0, a_1, \ldots, a_n]$ written in reduced form.

Finally, we show that every rational r can be expressed as a finite continued fraction. Set a_0 to be the largest integer less than or equal to r. If r is an integer, stop. (We have $r = [a_0]$.) Otherwise write

$$r = a_0 + \frac{1}{\varepsilon_0}$$

and let a_1 be the largest integer less than or equal to ε_0. If ε_0 is an integer, stop. (We have $r = [a_0, a_1]$.) Otherwise write

$$\varepsilon_0 = a_1 + \frac{1}{\varepsilon_1}$$

and continue. At each stage ε_i is a rational number greater than one. When written in reduced form, we see that the numerator of ε_{i+1} is the denominator ε_i, and since ε_{i+1} is greater than one, its denominator is smaller than this. Thus the numbers that appear as numerator and denominator in each fraction ε_i reduce in size, and so this process must eventually stop.

Challenge. i) Show that $[a_0, a_1, \ldots, a_n, 1] = [a_0, a_1, \ldots, a_n + 1]$. (So, as is customary we can assume that the final term in a finite continued fraction is not 1.)

ii) With this assumption, show that if $[a_0, a_1, \ldots, a_n] = [b_0, b_1, \ldots, b_m]$ with each a_i and each b_i a positive integer (except possibly $a_0 = 0$ or $b_0 = 0$), then $n = m$ and $a_0 = b_0, a_1 = b_1, \ldots, a_n = b_n$.

Being Clever

We can write the recursion relations

$$p_{n+1} = p_n a_{n+1} + p_{n-1},$$
$$q_{n+1} = q_n a_{n+1} + q_{n-1}$$

as a matrix product

$$\begin{pmatrix} p_{n+1} & p_n \\ q_{n+1} & q_n \end{pmatrix} = \begin{pmatrix} p_n & p_{n-1} \\ q_n & q_{n-1} \end{pmatrix} \cdot \begin{pmatrix} a_{n+1} & 1 \\ 1 & 0 \end{pmatrix}.$$

Thus

$$\begin{pmatrix} a_0 & 1 \\ 1 & 0 \end{pmatrix} \cdot \begin{pmatrix} a_1 & 1 \\ 1 & 0 \end{pmatrix} \cdot \ldots \cdot \begin{pmatrix} a_n & 1 \\ 1 & 0 \end{pmatrix} = \begin{pmatrix} p_n & p_{n-1} \\ q_n & q_{n-1} \end{pmatrix}.$$

Here $\frac{p_n}{q_n}$ (the first matrix column) is the fraction $[a_0, a_1, \ldots, a_{n-1}, a_n]$ and $\frac{p_{n-1}}{q_{n-1}}$ (the second matrix column) is $[a_0, a_1, \ldots, a_{n-1}]$.

Transposing gives

$$\begin{pmatrix} a_n & 1 \\ 1 & 0 \end{pmatrix} \cdot \begin{pmatrix} a_{n-1} & 1 \\ 1 & 0 \end{pmatrix} \cdot \ldots \cdot \begin{pmatrix} a_0 & 1 \\ 1 & 0 \end{pmatrix} = \begin{pmatrix} p_n & q_n \\ p_{n-1} & q_{n-1} \end{pmatrix}$$

so

$$[a_n, a_{n-1}, \ldots, a_1, a_0] = \frac{p_n}{p_{n-1}}$$

and

$$[a_n, a_{n-1}, \ldots, a_1] = \frac{q_n}{q_{n-1}}.$$

The first equation establishes the following curious fact:

Lemma 3. *If we reverse the terms of a finite continued fraction, we obtain a fraction with the same numerator as the original continued fraction.*

It is not obvious, for example, that the fractions

$$5003 + \cfrac{1}{2 + \cfrac{1}{7 + \cfrac{1}{418 + \cfrac{1}{92 + \cfrac{1}{3}}}}} \quad \text{and} \quad 3 + \cfrac{1}{92 + \cfrac{1}{418 + \cfrac{1}{7 + \cfrac{1}{2 + \cfrac{1}{5003}}}}}$$

should have the same numerator!

Theorem 4: *A palindromic continued fraction with an even number of terms has a numerator that is a sum of two squares.*

Consider the palindromic continued fraction $[a_0, a_1, \ldots, a_n, a_n, \ldots, a_1, a_0]$. The associated matrix product is

$$\begin{pmatrix} a_0 & 1 \\ 1 & 0 \end{pmatrix} \cdot \begin{pmatrix} a_1 & 1 \\ 1 & 0 \end{pmatrix} \cdots \begin{pmatrix} a_n & 1 \\ 1 & 0 \end{pmatrix} \cdot \begin{pmatrix} a_n & 1 \\ 1 & 0 \end{pmatrix} \cdots \begin{pmatrix} a_1 & 1 \\ 1 & 0 \end{pmatrix} \cdot \begin{pmatrix} a_0 & 1 \\ 1 & 0 \end{pmatrix}$$

$$= \begin{pmatrix} p_n & p_{n-1} \\ q_n & q_{n-1} \end{pmatrix} \cdot \begin{pmatrix} p_n & q_n \\ p_{n-1} & q_{n-1} \end{pmatrix} = \begin{pmatrix} p_n^2 + p_{n-1}^2 & p_n q_n + p_{n-1} q_{n-1} \\ p_n q_n + p_{n-1} q_{n-1} & q_n^2 + q_{n-1}^2 \end{pmatrix}.$$

This shows that the numerator of $[a_0, a_1, \ldots, a_n, a_n, \ldots, a_1, a_0]$ is $p_n^2 + p_{n-1}^2$, where p_n is the numerator of $[a_0, a_1, \ldots, a_n]$ and p_{n-1} is the numerator of $[a_0, a_1, \ldots, a_{n-1}]$.

Comment. For a palindromic continued fraction with an odd number of terms (greater than one), all we can say about its numerator is that it is composite:

$$\begin{pmatrix} a_0 & 1 \\ 1 & 0 \end{pmatrix} \cdot \begin{pmatrix} a_1 & 1 \\ 1 & 0 \end{pmatrix} \cdots \begin{pmatrix} a_n & 1 \\ 1 & 0 \end{pmatrix} \cdot \begin{pmatrix} b & 1 \\ 1 & 0 \end{pmatrix} \cdot \begin{pmatrix} a_n & 1 \\ 1 & 0 \end{pmatrix} \cdots \begin{pmatrix} a_1 & 1 \\ 1 & 0 \end{pmatrix} \cdot \begin{pmatrix} a_0 & 1 \\ 1 & 0 \end{pmatrix}$$

$$= \begin{pmatrix} p_n & p_{n-1} \\ q_n & q_{n-1} \end{pmatrix} \cdot \begin{pmatrix} b & 1 \\ 1 & 0 \end{pmatrix} \cdot \begin{pmatrix} p_n & q_n \\ p_{n-1} & q_{n-1} \end{pmatrix}$$

$$= \begin{pmatrix} p_n (b p_n + 2 p_{n-1}) & b p_n q_n + p_{n-1} q_n + p_n q_{n-1} \\ b p_n q_n + p_{n-1} q_n + p_n q_{n-1} & q_n (b q_n + 2 q_{n-1}) \end{pmatrix}$$

and $p_n (b p_n + 2 p_{n-1})$ is composite for $n \geq 2$ because the values of p_n grow in size: $a_0 = p_0 < p_1 < p_2 < \cdots$.

Making Use of this Sum of Two Squares

At this stage we invoke an inspired method of Art Benjamin and Doron Zeilberger ([BENJAMIN and ZEILBERGER]).

Theorem 5: *If p is a prime of the form $4n + 1$ for $n \geq 1$, then p is the sum of two squares.*

Proof. Consider the fractions: $\frac{p}{1}, \frac{p}{2}, \frac{p}{3}, \ldots, \frac{p}{2n}$, each a quantity greater than two.

Write $\frac{p}{k}$ as a continued fraction of the form $[a_0, a_1, \ldots, a_n]$ with $a_0 \geq 2$ (which will be the case since each fraction is greater than 2) and with $a_n \geq 2$ (which we can do because $[a_0, a_1, \ldots, a_{n-1}, 1] = [a_0, a_1, \ldots, a_{n-1} + 1]$). By lemma 3, its reverse, $[a_n, a_{n-1}, \ldots, a_0]$, is a fraction with the same numerator p. As it is also greater than two ($a_n \geq 2$), it must be another number in the list.

Notice, $\frac{p}{1} = [p]$ is "self dual." Since there are an even number of fractions among $\frac{p}{1}, \frac{p}{2}, \frac{p}{3}, \ldots, \frac{p}{2n}$, at least one other fraction is self-dual.

Thus we have a fraction with prime p numerator whose continued fraction expansion is palindromic. By the comment made after Theorem 4, the expansion cannot contain an odd number of terms. Consequently, there are an even number of terms and, by Theorem 4, the numerator p is a sum of two squares. □

Comment. No prime of the form $4n + 3$ is a sum of two squares. To see why, note that the square of an even number is a multiple of four and the square of an odd number is one greater than a multiple of four. Consequently, any number that is the sum of two squares can leave a remainder of only 0, 1, or 2 on division by four.

Sums of Two Squares: Some Congruence Results

We say "a is congruent to r modulo p" if a and r leave the same remainder on division by p. We write $a \equiv r \pmod{p}$.

Every a is congruent, modulo p, to exactly one of $0, 1, 2, \ldots, p - 1$. If p is prime, then it is not too hard to prove that for a non-zero a, the entries $0, 1a, 2a, \ldots, (p-1)a$ are distinct when taken modulo p. They thus represent a rearrangement of $0, 1, 2, \ldots, p - 1$. This leads to

Lemma 6. *If p is prime, given non-zero integers a and b, we can find an integer m so that $am \equiv b \pmod{p}$.*

If N is a sum of two squares, $N = x^2 + y^2$, and $d = \gcd(x, y)$, then set $a = \frac{x}{d}$ and $b = \frac{y}{d}$. Then $N = d^2(a^2 + b^2)$ with a and b coprime.

Suppose, in the prime factorization of N, a prime p appears an odd number of times. Then it appears at least once in the prime factorization of $a^2 + b^2$ and so $a^2 + b^2$ is divisible by p. This means that neither a nor b can be a multiple of p. (If $p \mid a$ and $p \mid (a^2 + b^2)$, then $p \mid b^2$ and hence $p \mid b$, and a and b are not coprime after all.)

Choose m and n so that

$$am \equiv b \pmod{p} \text{ and } an \equiv 1 \pmod{p}.$$

Since $a^2 + b^2$ is a multiple of p we have $a^2 \equiv -b^2 \pmod{p}$. Then

$$a^2 m^2 \equiv b^2 \equiv -a^2 \pmod{p}$$

so

$$n^2 a^2 m^2 \equiv -a^2 n^2 \pmod{p}.$$

That is,

$$1^2 \cdot m^2 \equiv -1^2 \pmod{p}$$

or

$$m^2 \equiv -1 \pmod{p}.$$

We will next prove

Theorem 7: *If p is a prime three greater than a multiple of four, then there is no integer m such that $m^2 \equiv -1 \pmod{p}$.*

Thus, if p is a prime three greater than a multiple of four appearing an odd number of times in the prime factorization of N, we have a contradiction. So we have

Corollary 8. *If a number is a sum of two squares, then any prime congruent to 3 modulo four that appears in its prime factorization does so an even number of times.*

Proving Theorem 7

We need to explore congruence results. Suppose p is a prime different from 2. (Systems modulo two are simple!) As we have seen, for any non-zero integer a, the list $1a, 2a, \ldots, (p-1)a$, modulo p, is just a rearrangement of the list $1, 2, \ldots, p-1$. Thus for each non-zero b, we can find a number m so that $b = ma \pmod{p}$ that is unique modulo p.

Working with $b = 1$

For each a there is an m so that $am \equiv 1 \pmod{p}$. We denote m as a^{-1} and call it the inverse of a. Notice that 1 and -1 are self-inverses (clearly $1 \cdot 1 \equiv 1 \pmod{p}$ and $(-1) \cdot (-1) \equiv 1 \pmod{p}$) and these are the only self-inverses because if $x^2 \equiv 1 \pmod{p}$ then $p \mid (x^2 - 1)$ and so $p \mid (x - 1)$ or $p \mid (x + 1)$ whence $x \equiv \pm 1 \pmod{p}$.

The product:

$$1 \cdot 2 \cdot \cdots \cdot (p - 1)$$

modulo p is the product of 1, of -1 (which is $p - 1$), and of pairs of the form x and x^{-1}, each producing a product of 1. Thus

$$(p - 1)! \equiv -1 \pmod{p},$$

which is Wilson's theorem.

Working with $b = -1$

For each a there is an m so that $am \equiv -1 (\text{mod } p)$. Call a number *self-dual* if $a^2 \equiv -1 (\text{mod } p)$.

For each prime p there are either two self-duals in the list $1, 2, \ldots, p - 1$, or none.

If $a^2 \equiv -1 (\text{mod } p)$ then $(-a)^2 \equiv -1 (\text{mod } p)$ as well (and, since p is odd, a and $-a$ are distinct modulo p). So if there is one self-dual there are at least two. Could there be a third self-dual b with $b^2 \equiv -1 (\text{mod } p)$? If so, then $a^2 - b^2 \equiv 0 (\text{mod } p)$ meaning that p is a factor of $a^2 - b^2 = (a - b)(a + b)$ so p is either a factor of $a - b$ (giving $b \equiv a (\text{mod } p)$) or a factor of $a + b$ (giving $b \equiv -a (\text{mod } p)$). There is thus no third self-dual.

If there are no self-duals, we have

$$(p - 1)! \equiv (-1)^{\frac{p-1}{2}} \ (\text{mod } p).$$

If there are two self duals a and $-a$ we have

$$\begin{aligned}(p - 1)! &\equiv a \cdot (-a) \cdot (-1)^{\frac{p-3}{2}} \ (\text{mod } p) \\ &\equiv -a^2 \cdot (-1)^{\frac{p-3}{2}} \ (\text{mod } p) \\ &\equiv -(-1) \cdot (-1)^{\frac{p-3}{2}} \ (\text{mod } p) \\ &\equiv (-1)^{\frac{p-3}{2}} \ (\text{mod } p).\end{aligned}$$

We know by Wilson's theorem that $(p - 1)! \equiv -1 (\text{mod } p)$. If p is of the form $4n + 3$, then $(-1)^{\frac{p-3}{2}} = (-1)^2 = 1$, which is impossible. This means that there are no self duals and hence there can be no m with $m^2 \equiv -1 (\text{mod } p)$.

This establishes Theorem 7.

Putting It All Together

We have:

- Each prime of the form $4n + 1$ is a sum of two squares.
- 2 is a sum of two squares ($1^2 + 1^2$).
- Any number raised to an even power is a sum of two squares ($a^{2n} = (a^n)^2 + 0^2$).
- The product of sums of two squares is a sum of two squares.
- If a number is a sum of two squares, then any prime of the form $4n + 3$ that appears in its prime factorization does so an even number of times.

These fit together to establish—at long last—the result:

An integer is a sum of two squares if and only if each prime of the form $4n + 3$ that appears in its prime factorization does so an even number of times.

Final Thought

Can anything be said about the count of ways we can express a number as a sum of two squares?

The number 10 can be written as a sum of two squares in, essentially, just one way, $10 = 3^2 + 1^2$, and the number 25 in essentially two different ways: $25 = 5^2 + 0^2 = 4^2 + 3^2$. (And numbers involving an odd power of the wrong type of prime in no ways!)

In 1800, at the age of 23, Carl Friedrich Gauss discovered a remarkable relationship between these counts and the number π. (See [HONSBERGER].) We will count the number of representations of a number as a sum of two squares of positive and negative integers and consider the order of the terms important. Thus 10 has eight representations:

$$
\begin{aligned}
10 &= 3^2 + 1^2 & &= 1^2 + 3^2 \\
&= 3^2 + (-1)^2 & &= (-1)^2 + 3^2 \\
&= (-3)^2 + 1^2 & &= 1^2 + (-3)^2 \\
&= (-3)^2 + (-1)^2 & &= (-1)^2 + (-3)^2
\end{aligned}
$$

and 25 has twelve representations.

Here's a table of the number of representations for each of the integers zero through ten:

Number	Count
0	1
1	4
2	4
3	0
4	4
5	8
6	0
7	0
8	4
9	4
10	8

The average value of the count is

$$
\frac{1+4+4+0+4+8+0+0+4+4+8}{11} \approx 3.36.
$$

What is the average value in the long run? That is, if $r(N)$ represents the number of ways we can write N as a sum of two squares, what is the value of

$$
\lim_{N \to \infty} \frac{r(0) + r(1) + r(2) + \cdots + r(N)}{N+1},
$$

the average number of ways to write a number as a sum of two squares?

Theorem. $\lim_{N \to \infty} \frac{r(0)+r(1)+r(2)+\cdots+r(N)}{N+1} = \pi.$

Proof. The sum $r(0) + r(1) + r(2) + \cdots + r(N)$ is the number of pairs of integers (a, b) with $a^2 + b^2 = k$ for some $k \le N$. That is, it is the number of pairs of integers (a, b) satisfying $a^2 + b^2 \le N$, the number of integer points inside a circle centered at the origin with radius $\sqrt{N}$.

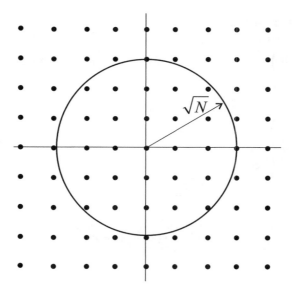

Surround each interior point by a unit square with vertical and horizontal sides at half-integer values. Their union approximates the interior of the circle.

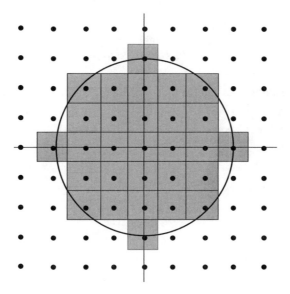

So we can approximate $r(0) + r(1) + r(2) + \cdots + r(N)$ by $\pi(\sqrt{N})^2 = \pi N$, the area of a circle of radius $\sqrt{N}$, and so

$$\lim_{N \to \infty} \frac{r(0) + r(1) + r(2) + \cdots + r(N)}{N + 1} \approx \lim_{N \to \infty} \frac{\pi N}{N + 1} = \pi.$$

We can make this rigorous by noting that no unit square extends beyond the circle of radius $\sqrt{N}$ by more than half the diagonal of the square (which has length $\frac{\sqrt{2}}{2}$) and no gap inside the circle extends inwards by more than $\frac{\sqrt{2}}{2}$ units. Thus

$$\pi \left(\sqrt{N} - \frac{\sqrt{2}}{2} \right)^2 \leq r(0) + r(1) + r(2) + \cdots + r(N) \leq \pi \left(\sqrt{N} + \frac{\sqrt{2}}{2} \right)^2.$$

Dividing by $N + 1$ gives

$$\pi \left(\sqrt{\frac{N}{N+1}} - \frac{\sqrt{2}}{2\sqrt{N+1}} \right)^2 \leq \frac{r(0) + r(1) + r(2) + \cdots + r(N)}{N+1} \leq \pi \left(\sqrt{\frac{N}{N+1}} + \frac{\sqrt{2}}{2\sqrt{N+1}} \right)^2$$

and taking the limit as $N \to \infty$ gives

$$\pi \left(1 - 0 \right) \leq \lim_{N \to \infty} \frac{r(0) + r(1) + r(2) + \cdots + r(N)}{N+1} \leq \pi \left(1 + 0 \right),$$

so

$$\lim_{N \to \infty} \frac{r(0) + r(1) + r(2) + \cdots + r(N)}{N+1} = \pi.$$

$\square$

Challenge. On average, in how many ways can a number be represented as a sum of three squares?

References

[BENJAMIN and ZEILBERGER] Benjamin, A., and Zeilberger, D. , Pythagorean primes and palin-dromic continued fractions, *Integers: Electronic Journal of Combinatorial Number Theory*, **5** (1) (2005), #A30.

[HARDY and WRIGHT] Hardy, G. H., and Wright, E. M., *An Introduction to the Theory of Numbers*, Clarendon Press, Oxford, England, 1979.

[HONSBERGER] Honsberger, R., *Ingenuity in Mathematics*, Mathematical Association of America, Washington D.C., 1970.

Appendix II
Pick's Theorem

Pick's Theorem—and Beyond!

Middle and high school students attending the spring 2009 St. Mark's Institute of Mathematics research class explored Pick's formula and developed an innovative approach to explaining it. It was an honor for me to witness the beautiful story they discovered unfold!

A version of this article appeared in the February/March 2010 issue of *FOCUS*, the news magazine of the Mathematical Association of America:

The Mathematical Artists

Chiron Anderson, Shivani Angappan, Adam Cimpeanu, Kevin Dibble, Swetha Dravida, Theo Fitzgerald, Bianca Homberg, Steven Homberg, Eric Marriott, Curtis Mogren, Hilary Mulholland, Alexandra Palocz, Linus Schultz, William Sherman, Alex Smith, David Tang, Steven Tang, and Andrew Ward.

The Article

In 1899 George Pick discovered a relationship between the area A of a simple lattice polygon (that is, one whose vertices have integer coordinates), the number b of lattice points on the boundary of the polygon, and the number i of lattice points inside the polygon. He showed that

$$A = i + \frac{b}{2} - 1.$$

The formula is usually proved by subdividing the polygon into lattice triangles after having verified that the result is true for all lattice triangles and that it remains true if a triangle is attached to a polygon for which the formula already holds. The greatest difficulty lies in establishing the result for triangles.

In the Spring of 2009 young students (ages 9–17) attending the St. Mark's Institute of Mathematics research group took another approach to Pick's theorem. They began by questioning the coefficients that appear in the formula: Why are interior points each worth 1? Why are boundary points each worth $1/2$? Examination of a lattice rectangle leads to an insight.

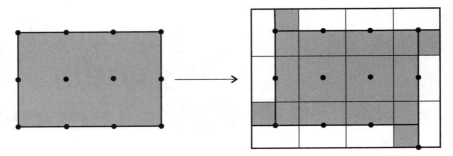

Figure 1

If we surround each lattice point with a unit square with sides at half-integer coordinates (let's call these squares *cells*) then we see that each interior point contributes one full square unit of area and each boundary point different from a vertex half a unit of area. If we extend the sides of the rectangle to make its exterior angles explicit we can introduce additional area so that each vertex also contributes half a unit of area. As the exterior angles of any polygon sum to one full turn, this excess in area amounts to one full square unit. The -1 in Pick's formula compensates for this.

This applies directly to any simple polygon with sides parallel to the axes of the lattice provided exterior angles turned in opposite directions are deemed opposite in sign.

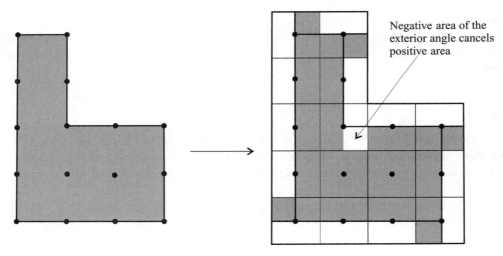

Negative area of the exterior angle cancels positive area

Figure 2

And essentially the same argument applies to any simple lattice polygon! The key is to note that diagonal line segments connecting two lattice points are rotationally symmetric about their midpoints. That is, any cell that is intercepted by a diagonal and divided into two parts is matched by a rotationally symmetric cell divided into the same two parts. (The matching portions are on alternate sides of the diagonal.)

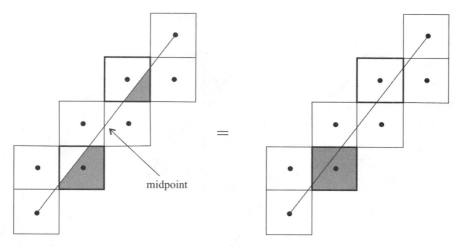

Figure 3

Each subdivided cell whose center is an interior point of the polygon can thus be completed by switching rotationally symmetric portions. (The analogous result holds for each cell with center in the exterior of the polygon.) We are close to a Proof Without Words of Pick's result:

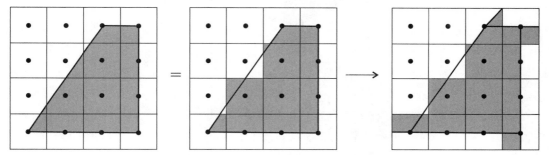

Figure 4: Each interior point contributes one unit of area and each boundary point half a unit of area with an overall error of -1.

A complication arises when more than one diagonal passes through the same cell. This can be handled by switching portions about one diagonal at a time. First choose a portion within a cell on one side of a diagonal that is free from other intercepting diagonals. If its rotationally symmetric counterpart is also free from intercepting diagonals, perform the switch. If not, work with a smaller part of the matching cell and attempt a switch there. As there are only finitely many regions to consider, there is sure to be a first switch to perform and all maneuvers thereafter will fall into place.

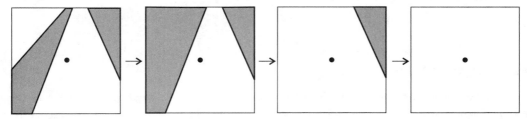

Figure 5: A series of switches.

This technique shows, for example, that the area of the following triangle is one half:

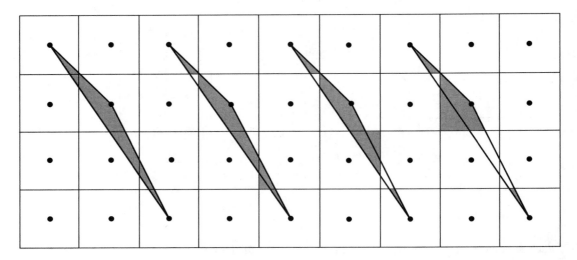

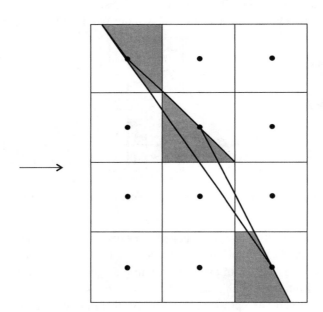

Figure 6: Pick's Theorem for a lattice triangle lacking interior points.

We have established Pick's theorem.

Generalizing the Result

Of course youngsters want to take a result and twist it in new and intriguing ways. (Just as do mathematicians!) Does a version of Pick's theorem hold for polygons with holes? For polygons with tendrils? For disconnected shapes?

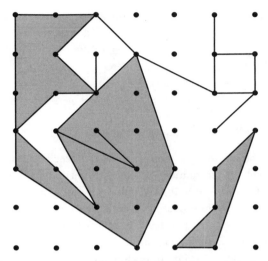

Figure 7: A wild "polygon." Shaded regions indicate area to be evaluated.

As the ordinary Pick's theorem assigns weights to points of the polygon (a weight of 1 for interior points and ½ for boundary points), students developed a general system for assigning weights to points of any shape. First take note of the shading. If an edge has the same coloring on both of its sides, thicken it slightly and produce either an infinitesimal amount of space or an infinitesimal amount of area wedged between two infinitesimally close edges.

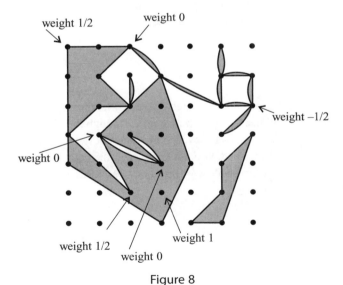

Figure 8

For each lattice point considered part of the shape assign a weight of $1 - \frac{e}{4}$ where e is the number of edges emanating from that point. Every other lattice point in the plane is given weight zero. Let w be the total sum of all the weights. For the shape in Figure 8, $w = 11\frac{1}{2}$.

Let h be the number of holes that appear in the thickened shape (including the infinitesimally small holes), c the number of components of the shape, and A the area of the shaded regions (disregarding infinitesimal regions, that is, A is the area of the original unthickened shape). In our example, $h = 5$, $c = 2$ and $A = 14\frac{1}{2}$. Students discovered

Generalized Pick's Theorem: $A = w + h - c$.

For a simple polygon, $w = i + b/2$, $h = 0$ and $c = 1$ and we have Pick's statement.

Outline of proof: Consider the original polygon (with no thickened edges) with shading removed. Using an induction argument on the number of edges connecting a fixed set of vertices, it is straightforward to show that $w + h - c = 0$. (We must still consider thickened edges to correctly determine weights.) This is just the Euler-Descartes formula in disguise. Draw in additional edges to connect the components of the graph (so that c becomes 1) and to make each region a simple polygon. We still have $w + h - c = 0$. Now return the shading of the original shape one region at a time. A little thought shows that the weight of each boundary point of that region (suppose there are b of them) increases by $1/2$ and the weight of each interior point of that region (say there are i of them) increases by 1. The number of holes of the modified shape decreases by 1. Thus we have:

$$A \rightarrow A + \text{area of new shaded region}$$
$$w \rightarrow w + i + b/2,$$
$$h \rightarrow h - 1 \quad,$$
$$c \rightarrow c.$$

By Pick's original theorem, the formula $A = w + h - c$ thus remains true at each step. At the end of the process remove the additional edges one at a time. This too does not disrupt the validity of the formula.

Conclusion

What clever youngsters! The extracurricular research classes I offer through the St. Mark's Institute of Mathematics are conducted as a conversational Math Circle, through the give-and-take of discussion and exploration. My role in these sessions is never to instruct, but rather to guide, nudge, and offer my opinions and hunches only when asked. I do not know what the outcomes will be. To learn more about Math Circles, visit the MAA website www.maa.org and click on SIGMAA MCST (Math Circles for Students and Teachers).

Appendix III

The Möbius Function

The following article appeared in the March 2007 issue of *FOCUS* (volume 27, number 3), the news magazine of the Mathematical Association of America. It represents another testimony to the inspiration working with young students can provide.

An Illuminating Introduction to the Möbius Function

The Möbius function, for the purposes of the Möbius inversion formula, can be difficult to motivate in a first-experience number theory course. After teaching an extra-curricular class to motivated high-school students I was surprised to find a natural appearance of it (and motivation for inversion) through a classic puzzle, the starting point of our course:

> *Along a school corridor stand one hundred lockers, numbered 1 through 100, each initially closed. One hundred students, also numbered 1 through 100, take turns walking down the corridor. Student 1 opens every locker. Student 2 touches every second locker (lockers 2, 4, 6, ...) closing each. Student 3 touches every third locker, changing its state to closed if it was open, to open if it was closed, and so on, all the way down until the 100th student walks down and changes the state of every 100th locker (namely, just the final one!) After all one hundred students have walked, which lockers are left open?*

Students enjoy enacting a smaller version of this puzzle (use 25 cups or playing cards) and it is usually a surprise to all to find lockers 1, 4, 9, 16, 25, those with square numbers left open. It is not difficult to explain why this is the case:

> *Locker N is touched by student d only if $d \backslash N$. As only the square numbers possess an odd number of factors, the square numbered lockers are left open.*

The number of lockers specified in the problem is immaterial. For the sake of convenience let's assume the number of lockers, and the number of students, is infinite and in one-to-one correspondence with the set of natural numbers. (For the case of a finite number of lockers truncate the results that follow.)

I based my student course on a wonderful paper by Torrens and Wagon ([3]) that explores the possibility of sending down a subset of students to obtain a pre-described configuration of open and closed lockers. A first challenge of this type would ask: Which students should be sent down the corridor to set locker 1 open and the others closed? Considering the states of lockers 1, 2, 3, ... in turn leads to the subset of students $S = \{1, 2, 3, 5, 6, 7, 10, 11, 13, 14, 15, 17, \ldots\}$. This looks like the set of square-free numbers and it is not difficult to prove that this is indeed so because each number greater than one possesses an even number of square-free factors. Establishing this, and exploring other issues raised in [3], provided great material for a short course for beginning students. I was surprised to discover that a natural extension of the locker problem—also contemplated by my students—provides motivation for standard material of a college number-theory course.

Enter the Möbius Function

Lockers come in two states—open or closed—and alternate between them when touched. Let's now consider objects that cycle through k different states when touched for some $k \geq 2$. We represent the cycle of states as $0 \to 1 \to 2 \to \cdots \to k - 1 \to 0$ (or, more compactly, $i \to i + 1 \,(\text{mod } k)$). The objects could be light bulbs operated by simple pull strings. For $k = 4$, a bulb could cycle through the states off, dim, bright, and very bright. Consider the problem

> *Light bulbs numbered 1, 2, 3, ..., all initially off (state 0), line a corridor. A subset of students from a set of students numbered 1, 2, 3, ... shall be sent down the corridor. Student r, if in the subset, will pull the cord of each bulb whose number is a multiple of r. Which students should be sent down the corridor so as to set bulb 1 into state 1 and leave all the others in state 0? (Students may make repeat trips.)*

For $k = 4$, thinking about bulbs 1, 2, 3, ... in turn leads to the multi-subset of students:

$$S = \{1, 2, 2, 2, 3, 3, 3, 5, 5, 5, 6, 7, 7, 7, 10, 11, 11, 11, 13, 13, 13, 14, 15, 17, 17, 17, \ldots\},$$

where the number of times student r is listed in S corresponds to the number of times we must send student r down the corridor. (It is worth computing this set up to student 30, the first student whose number is a product of three distinct primes.)

Let s_n denote the number of times we must send student n down the corridor. Working mod k it appears that

$$s_n = \begin{cases} 1 & \text{if } n \text{ is the product of an even number of distinct primes} \\ -1 & \text{if } n \text{ is the product of an odd number of distinct primes} \\ 0 & \text{otherwise.} \end{cases}$$

(Here, sending a student down the corridor "-1 times" means sending that student down $k - 1$ times.) We have discovered the Möbius function: $s_n = \mu(n)$ (at least in a mod k setting). We need to prove that this multi-set of students really does do the trick.

Claim. *If student n is sent down the corridor $\mu(n)$ times, then bulb 1 will be in state 1 and all other bulbs in state 0.*

Proof. Bulb 1 will be touched only once and so will be in state 1. For $n > 1$, bulb n will be touched s_d times by student d for $d|n$ and by no other students. If n has prime factorization $n = p_1^{\alpha_1} p_2^{\alpha_2} \cdots p_t^{\alpha_k}$, then any factor d of n for which s_d is non-zero is a product of distinct primes from $\{p_1, p_2, \ldots, p_t\}$. There are $\binom{t}{r}$ factors d that are a product of r primes (for $1 \leq r \leq t$), and since $\binom{t}{0} - \binom{t}{1} + \binom{t}{2} - \cdots \pm \binom{t}{t} = 0$, it follows that $\sum_{d|n} s_d = 0$, and so bulb n will be in state 0. $\qquad\square$

Comment. We defined μ in a mod k setting, but it can defined as a function from $\mathbb{N}$ to $\mathbb{Z}$ with the same formula. The proof can be repeated, essentially verbatim, to establish the classic result

$\sum_{d|n} \mu(d) = \begin{cases} 0 & \text{if } n \neq 1 \\ 1 & \text{if } n = 1 \end{cases}$. It is also clear from its definition that $\mu(n)$ is a multiplicative function— another classic observation.

Enter the Möbius Inversion Formula

Rather than require just bulb 1 to be in state 1 and all other bulbs be in state 0 suppose, for each $n \in \mathbb{N}$, we desire bulb n to be in state b_n for $0 \leq b_n \leq k - 1$. This gives the sequence of bulb states:

$$B = \{b_1, b_2, b_3, \ldots\}.$$

Let s_n denote the number of times (mod k) that student n must be sent down to accomplish this configuration. It is clear we must set:

$$s_1 = b_1 \quad .$$

For bulb 2, we must send student 2 down the corridor enough times to counteract the effect of student 1, and then b_2 more times to achieve the desired state. Thus

$$s_2 = -s_1 + b_2 = -b_1 + b_2.$$

(This is to be interpreted mod k.)

In the same way we must have

$$s_3 = -s_1 + b_3 = -b_1 + b_3.$$

For bulb 4, we must counteract the effects of students 1 and 2 and then pull the cord of bulb 4 b_4 more times. This gives

$$s_4 = -s_1 - s_2 + b_4 = -b_2 + b_4.$$

Similarly,

$$s_5 = -s_1 + b_5 = -b_1 + b_5,$$
$$s_6 = -s_1 - s_2 - s_3 + b_6 = b_1 - b_2 - b_3 + b_6,$$
$$\ldots$$
$$s_{30} = -b_1 + b_2 + b_3 + b_5 - b_6 - b_{10} - b_{15} + b_{30}.$$

An astute student (and in my course there were several) might observe that this indicates the appearance of the Möbius function. We conjecture

$$s_n = \sum_{pq=n} b_p \mu(q).$$

Claim. *Given a sequence $B = \{b_n\}$ of bulb states, sending student n down $s_n = \sum_{pq=n} b_p \mu(q)$ times produces sequence B.*

Proof. Bulb n is touched s_d times by student d for $d|n$ and by no other student. Thus bulb n will be in state

$$\sum_{d|n} s_d = \sum_{d|n} \sum_{pq=d} b_p \mu(q) = \sum_{pqr=n} b_p \mu(q) = \sum_{p|n} b_p \sum_{d|\frac{n}{p}} \mu(d) = b_n.$$

$\square$

Conversely, given a student list:

$$S = \{s_1, s_2, s_3, \ldots\}$$

indicating the number of times student n will be sent down the corridor (mod k) the resulting state of bulb n will be

$$b_n = \sum_{d|n} s_d.$$

We have, in fact, established the Möbius inversion formula (mod k)

$$b_n = \sum_{d|n} s_d \quad \Leftrightarrow \quad s_n = \sum_{d|n} b_d \mu\left(\frac{n}{d}\right).$$

The proof of the Möbius inversion formula in general follows verbatim if we choose a k larger than any of the values $b_1, b_2, \ldots, b_n, s_1, s_2, \ldots, s_n$ for a fixed value n.

Final Comments

Letting go of our mod k thinking, the Möbius function appears in the problem of finding multiplicative inverses of Dirichlet series. (See [2] and [1] for instance.) For example, students can now prove:

$$\left(\sum_{n=1}^{\infty} \frac{1}{n^2}\right) \left(\sum_{m=1}^{\infty} \frac{\mu(m)}{m^2}\right) = 1.$$

The quantity $\sum_{m=1}^{\infty} \frac{\mu(m)}{m^2}$ can be interpreted as the probability of selecting two natural numbers at random that are relatively prime. (See [1].) It has value $\frac{6}{\pi^2}$.

References

Bridger, M.and Zelevinsky, A., Visibles revisited, *College Mathematics Journal,* **36** (2005), 269–300.

Cuoco,A., Searching for Möbius, *College Mathematics Journal,* **37** (2006), 137–142.

Torrence, B. and Wagon, S., The locker problem, *Crux Mathematicorum,* **34** (4) (2007), 232–236.

Appendix **IV**
The Borsuk-Ulam Theorem

Ever since I first learned of the Borsuk-Ulam theorem as an undergraduate I wondered if there was a straightforward approach to understanding why it was true. It wasn't until I witnessed the work of youngsters in the Spring of 2008 that I saw it. The following article appeared in the November 2008 issue of FOCUS (volume 28, number 8), the news magazine of the Mathematical Association of America. The proof was also the topic of the October 2008 St. Mark's Institute of Mathematics newsletter.

An Intuitive Approach to the Borsuk-Ulam Theorem

By Alex Bishop, Adam Cimpeanu, Kyle Flood, Bianca Homberg, Steven Homberg, Eric Marriott, Jeffrey Roth, Linus Schultz, William Sherman, Alex Smith, Geoffrey Smith and *James Tanton of the St. Mark's Institute of Mathematics.*

Each semester I offer extracurricular mathematics courses for math-interested students, ages 11–18, keen to experience mathematics as a creative and organic enterprise. Called research classes, they introduce students to the joys and frustrations of the research experience, of not knowing, of feeling around in the dark, and of finding the fortitude of mind to concentrate on complex issues for sustained periods. (Weeks, not just minutes.) This semester the coauthors of this paper established the classic Borsuk-Ulam Theorem in a way that is slick, intuitive, and accessible. Their success speaks to the joy and value that can be found for all in allowing mathematical creativity to bubble forth, no matter the age and the background experience of the mathematical artists!

The Theorem

Informally, the Borsuk-Ulam theorem states:

> *At any instant there exist two antipodal points on the Earth's surface with identical air temperature and air pressure.*

Here we are assuming that air temperature and air pressure vary continuously over the surface of the Earth. A more formal statement of the theorem would be that any continuous map $f : S^2 \to \mathbb{R}^2$ identifies a pair of antipodal points, but the interpretation with $f = ($temperature, pressure$)$ is usually offered in texts as an appealing and quirky consequence.

Here is the students' proof:

For a point P on the surface of the Earth let P^* denote its antipodal point (and so $P^{**} = P$), and for a continuous temperature/pressure map $f = (f_1, f_2) : S^2 \to \mathbb{R}^2$ consider the map $g : S^2 \to \mathbb{R}^2$ given by the differences of temperatures and pressures for pairs of antipodal points:

$$g(P) = (f_1(P) - f_1(P^*), \ f_2(P) - f_2(P^*)).$$

For any point P we have that $g(P^*) = -g(P)$, so if we plot the set of values of $g(P)$ in the plane $\mathbb{R}^2$ we obtain a set that is rotationally symmetric $180°$ about the origin.

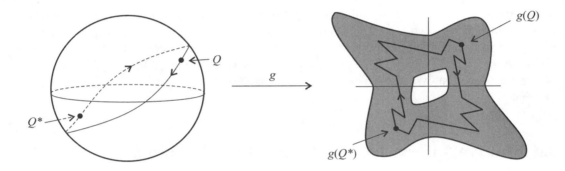

Pick a point Q on the sphere and walk along a great circle to its antipode Q^*. This gives a path in the image set. Walking back from Q^* to Q along the second half of the great circle gives the rotational image of this path in the image set. We now have a loop in the image set starting and ending at $g(Q)$.

Holding Q fixed, shrink (that is, contract) the great circle over the surface of the sphere down to the point Q. The image loop starting and ending at $g(Q)$ shrinks to the point $g(Q)$. Some intermediate loop passes through the origin so, despite what the diagram suggests, there is a point R for which $g(R) = (0, 0)$.

Down a Dimension

The one-dimensional version of the theorem states that for a continuous map on a circle there exist antipodal points adopting the same value. As my students discovered, this can be generalized. Their approach is novel.

For any continuous map $f : S^1 \to \mathbb{R}$ and any k there exists θ such that $f(\theta) = f(\theta + k)$.

(A real value θ corresponds to the point $(\cos\theta, \sin\theta)$ on the unit circle.)

To see why this is true, think of the function f as the height function for a fence built along a circle and let θ^* correspond to the location of the maximum height of the fence. Attach a rod to the top of the fence at position θ^* and at position $\theta^* + k$.

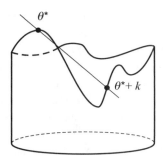

Slide the rod around the fence, keeping the pivot points of the rod at positions θ and $\theta + k$ until the second end of the rod reaches position θ^*. The slope of the rod has changed sign and so the rod must, at some position, have been horizontal.

Question. Is there a two-dimensional analog of this result?

Ham-Sandwiches

A famous consequence of the (two-dimensional) Borsuk-Ulam theorem states, somewhat informally

> *At any instant there exists a plane that divides the volume of Pluto, of the Eiffel Tower, and of this book in half simultaneously.*

This result is known as the ham-sandwich theorem for it can be phrased in terms of slicing two slices of bread and a slab of ham simultaneously in half.

To see why this is true, imagine that we have three sets S_1, S_2 and S_3 in space, each with a well-defined volume. A point P on a sphere with center O determines a vector $\overrightarrow{OP}$ and there exists a plane with $\overrightarrow{OP}$ as its normal that divides the volume of S_3 in half. Let $f_1(P)$ be the volume of the portion of S_1 that lies on the side of the plane that contains $\overrightarrow{OP}$ and $f_2(P)$ the volume of S_2 on this same side. We have defined a function $f : S^2 \to \mathbb{R}^2$ and, by the Borsuk-Ulam theorem, there exists a point R such that $f(R) = f(R^*)$. The planes corresponding to the points R and R^* are the same but with opposite orientation, so it is the one that does the slicing trick.

Although this argument for proving the ham-sandwich theorem is the standard one, it was also developed by the students.

A Context for Technicalities

To experienced mathematicians it is clear that many fundamental properties of continuity were used here: the intermediate-value theorem, the extreme-value theorem, an interplay between connectedness, path connectedness and simple connectedness, the use of continuous images of connected sets, of compact sets and the like. And many of these issues were up for discussion during the course. For example, in exploring the Borsuk-Ulam theorem students at first contemplated the image set $f\left(S^2\right)$ and wondered if it could have holes? In math-speak, they were asking: Can the continuous image of a simply connected compact set fail to be simply connected? If this were a college course I would have been grateful for the opportunity to leap in and begin a speech on beginning homotopy theory: a question has just provided a context and a need for a theory.

It is easy for an educator to fall into the trap of answering questions that have not been asked. Playing with ideas first and setting technicalities aside can provide material for questions and give context and meaning to technicalities that, in the end, cannot be ignored. Having a context is key to developing understanding and I hope as an educator that I can help students create a context. My young high-schoolers are now ready for a course on algebraic topology. Given that they are now also wondering what continuity means and whether it is possible to say that a fence will always have, somewhere, a maximum height they are ready for a course on real analysis too! I hope never to underestimate the value of intellectual play.

(By the way, the students achieved the results outlined in this paper in four one-hour sessions. The first four sessions of the semester were devoted to studying Sperner's lemma. My hunch was that it could be used to prove the Borsuk-Ulam theorem, but my hunch, as it turned out, was irrelevant!)

Appendix V
The Galilean Ratios

In the Spring of 2009, St. Mark's Institute students explored the charm of "proofs without words" and were inspired to come up with some astounding visual proofs of their own. I mentioned this in newsletter 10. This work was also printed in the October/November 2009 issue of FOCUS (volume 29, number 5), the news magazine of the Mathematical Association of America. That article is reprinted here.

Young Students Explore Proofs Without Words

Chiron Anderson, Shivani Angappan, Adam Cimpeanu, Kevin Dibble, Theo Fitzgerald, Bianca Homberg, Steven Homberg, Eric Marriott, Curtis Mogren, Alexandra Palocz, Linus Schultz, William Sherman, Alex Smith, David Tang, Steven Tang, Andrew Ward, and James Tanton of the St. Mark's Institute of Mathematics.

Looking for a discussion topic for math club? A tidbit to fill an idle moment in class? Try presenting a Proof Without Words – or two or three – as published in the journals of the MAA. You might be surprised by the conversations that follow!

In the spring of 2009 I presented young students (ages 9 – 17) of the St. Mark's Institute of Mathematics some wordless demonstrations of the Galilean ratios:

$$\frac{1}{3} = \frac{1+3}{5+7} = \frac{1+3+5}{7+9+11} = \frac{1+3+5+7}{9+11+13+15} = \cdots$$

Within minutes these youngsters came up with their own Proof Without Words of the result:

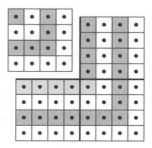

Figure 1

$$\frac{\text{The sum of the first } k \text{ odds}}{\text{The sum of next } k \text{ odds}} = \frac{1}{3}.$$

That was only the beginning!

Generalization 1. Students noted that a square also divides into 9 parts, 16 parts, and so on, and we can say more generally

$$\frac{\text{The sum of the first } k \text{ odds}}{\text{The sum of next } mk \text{ odds}} = \frac{1}{(m+1)^2 - 1}.$$

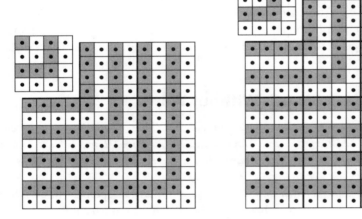

Figure 2

For example,

$$\frac{1}{3+5} = \frac{1+3}{(5+7)+(9+11)} = \frac{1+3+5}{(7+9+11)+(13+15+17)} = \cdots = \frac{1}{8}$$

and

$$\frac{1}{3+5+7} = \frac{1+3}{(5+7)+(9+11)+(13+15)} = \cdots = \frac{1}{15}.$$

Generalization 2. The choice of coloring was fortuitous and inspired more. If we adjust the shading of squares of odd dimension and regard the gray cells as positive and white cells as negative, we discover an alternating version of the Galilean ratios.

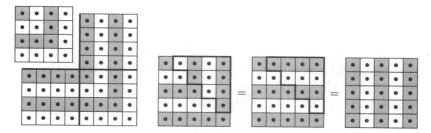

Figure 3

$$\frac{1}{3} = \frac{1-3+5}{7-9+11} = \frac{1-3+5-7+9}{11-13+15-17+19} = \cdots$$

Alternatively, if we regard the grey squares as worth a and the white squares as b we obtain

$$\frac{a}{3a} = \frac{a+3b+5a}{7a+9b+11a} = \frac{a+3b+5a+7b+9a}{11a+13b+15a+17b+19a} = \cdots$$

[What results follow from using more than two colors?]

Bonus. Our side picture for odd squares establishes $\frac{3}{1+5} = \frac{1}{2}$, $\frac{3+7}{1+5+9} = \frac{2}{3}$, $\frac{3+7+11}{1+5+9+13} = \frac{3}{4}$, and, in general, $\frac{3+7+\cdots+(4n-1)}{1+5+\cdots+(4n+1)} = \frac{n}{n+1}$.

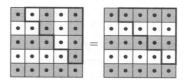

Figure 4

Generalization 3. A path towards more results is clear. Pictures like

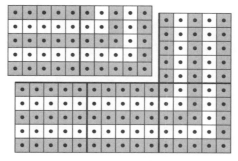

Figure 5

show, for instance, that

$$\frac{2}{4} = \frac{3+5}{7+9} = \frac{4+6+8}{10+12+14} = \frac{5+7+9+11}{13+15+17+19} = \frac{6+8+10+12+14}{16+18+20+22+24} = \cdots.$$

Increasing the number of rows and columns in this picture (as in generalization 1) yield many similar results.

If we reduce the fractions with even terms we obtain

$$\frac{1}{2} = \frac{2+3+4}{5+6+7} = \frac{3+4+5+6+7}{8+9+10+11+12} = \cdots$$

which possesses its own Proofs Without Words:

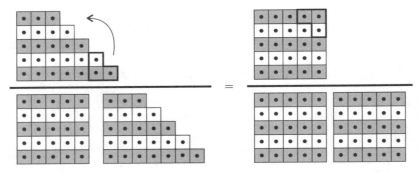

Figure 6

We see an alternating version of the result if we regard grey and white cells as opposite in parity. [Add rows of blocks to the diagram. What more can be discovered?]

Concluding Comments. This story of pictures took place over just a few forty-minute sessions and demonstrates the depth that can be attained from intellectual play. From my experience students of all ages are keen to engage in play when given the chance! Proofs Without Words provide a comfortable and appealing gateway to original enquiry.

Appendix VI
A Candy-Sharing Game

In the Spring of 2010, the Math Institute sponsored: The Great St. Mark's Institute of Mathematics Research Extravaganza. It was an e-mail challenge inviting followers of the Institute to explore an open-ended question over a period of months. I promised electronic gift certificates to those who I deemed had made significant progress with the project by the end of May. (The challenge turned out to be harder than I realized! I ended up being very generous and supplied certificates to all those who made a serious attempt.)

Here is the challenge and some of the results I know about it.

The Great Candy-Sharing Game

The Set-up. *A group of students sit in a circle with a number of pieces of candy randomly distributed among them. They agree to take part in a sharing game. At the blow of a whistle any student with two or more pieces of candy will give two pieces away: one to her left neighbor and one to her right neighbor. (If a neighbor also possesses two or more pieces of candy, then this pair will be simultaneously giving and receiving pieces of candy.) Students regroup their pieces of candy and repeat the process at the next blow of the whistle. The game is repeated until no one can make a move.*

Example. Here's a game with five students playing with five pieces of candy.

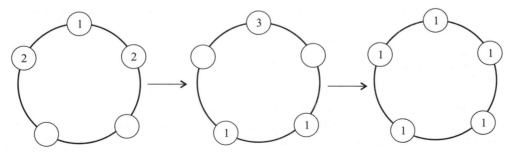

The game terminates after two turns.

Make sure you understand the play of the game here.

Practice. Here's another game with five students and five pieces of candy. Play it to see that it does not terminate.

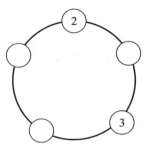

The research project is to explore these candy games! Here are some of the questions you might wish to think about.

a)   [The answer is known.] **Explain why a game with more candy than students will never terminate.**

b)   [The answer is known.] **Does the following game among eight players with eight pieces of candy terminate?**

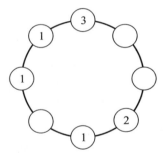

c)   **Consider a game with three students and three pieces of candy. List the possible distributions that lead to games that terminate. Can you do the same for a game of four students with four pieces of candy?**

d)   [The answer is known ... but can you come up with a simple way of answering it? Dr. T's approach is pretty hard!] **Must a game with fewer pieces of candy than students terminate?**

For example, if five pieces of candy are shared among six students, is the game sure to stop? How about ten players with nine pieces of candy?

e)   [The answer is known.] **In a game that terminates, must players share their candy simultaneously? Or could players take turns sharing two pieces? Does the order in which players take turns affect the outcome of a game?**

Try this too! Take a game that you know terminates. Play it several different times taking different turns as to which player acts when. Is the final outcome always the same?

f) [OPEN RESEARCH] *No one currently knows the answer to this question!* **Is there a way to determine whether or not a distribution of N pieces of candy among N students will lead to a terminating game without playing the game? Do terminating distributions share a recognizable common property?**

g) **[[OPEN RESEARCH] Suppose c pieces of candy are distributed among N players.**

Question d) asks about $c < N$.

Question f) observes that $c = N$ is complicated.

Question a) observes that the game will never terminate for $c > N$.

The case $c > N$ seems to have some curious structure. For example, a colleague has observed that for $N < c < 2N$, although the game does not terminate, the pattern of candy distribution seems to always oscillate between two states.
  For $2N < c < 3N$, does the game eventually oscillate between three states? What about for $3N < c < 4N$?
  Then there are the cases $c = 2N, c = 3N$, and so on.

Comment. Questions e), f), and g) suggest that it would be worth collecting data. Play candy games in different ways and see if any patterns emerge. Tabulating the behavior games would be a significant contribution to understanding.

Here are my solutions to the parts labeled as known. Just to be clear, we define "terminate" to mean "reach a state in which no player has two or more pieces of candy" (that is, there no longer is a player who can share).

Observation 1. *No game with more candy than students can terminate.*
  In such a game there will always be at least one player with two or more pieces of candy and so there will be another play to be made. (If each player has two or more pieces of candy, the distribution pattern of the candy does not change, but the game goes on.)

Observation 2. *In a game with an even number of students, if the total amount of candy possessed by every second student is odd (even), then that total remains odd (even) throughout the entire play of the game.*
  Give a hat to every second student and consider the total amount of candy the hatted students possess. If a hatted student gives away two pieces of candy, the total count decreases by two. If a non-hatted student gives away two pieces, the count increases by two. The parity of the total count therefore does not change.
  Thus it is impossible then for the following game with eight pieces of candy distributed among eight students to terminate with one piece of candy per student.

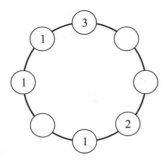

Observation 3. *A game with less candy than students must terminate.*

This is tricky to establish. My thanks to Stan Wagon for an e-mail interchange that helped get the following proof.

Proof of Observation 3. Suppose we have a distribution of less than N pieces of candy among N players sitting in a circle. We need some terminology.

Let's call a string of k consecutive players ($1 \leq k \leq N - 1$) *balanced* if they possess exactly k pieces of candy among them. Call the string *abundant* if they possess more than k pieces in total, and *deficient* if they have less than k pieces. For a deficient string, let's define its *deficiency* to be the number of additional pieces of candy they need to make the string balanced. (Thus a string of k students possessing $k - a$ candy pieces has deficiency a.) We'll say the deficiency of a balanced string and of an abundant string is zero.

For any distribution of candy among the N players, define the *index* of the distribution to be the sum of the deficiencies of all possible strings of lengths 1 through $N - 1$ among the students.

Since there is less candy than students, there is at least one student with no candy. This student represents a deficient string of length 1 and deficiency 1. Thus the index of any distribution is a positive integer.

Now ... At the blow of the whistle, those students with two or more pieces of candy will simultaneously give away two pieces of candy. To make the following analysis a little simpler, assume that they share their candy in turn one at a time.

We claim: *Each time a student shares two pieces of candy, the index of the distribution decreases.*

As the index is a positive integer, it cannot decrease indefinitely. Thus, if the claim is true, the candy-sharing game reaches a distribution for which no more sharing can occur. That is, the game must terminate.

Now to prove the claim:

Suppose Bettina is the student sharing two pieces of candy. Suppose she sits between Albert and Cuthbert.

Bettina represents her own (abundant) string of length 1. After sharing her candy, Bettina might become deficient, resulting in a contribution of $+1$ to the index. If she does not become deficient, she alone does not change the index.

The set of all students but Bettina constitutes a string of length $N - 1$. Since Bettina possesses at least two pieces of candy and there are no more than $N - 1$ pieces in all, the string is deficient by at least two. Bettina's sharing of candy decreases the deficiency by two, changing the index by -2.

These two strings, as a pair, decrease the value of the index.

Now we examine other strings of students. The only strings that remain affected by Bettina's sharing are those that contain Albert or contain Cuthbert, but not both. Let's focus on those strings that contain Albert only. (The analysis for Cuthbert is the same.)

Let $[\ldots A]$ represent a string that ends with Albert, and $[\ldots AB]$ the same string with Bettina adjoined. When Bettina shares two pieces of candy, the students in $[\ldots A]$ gain a candy piece, and the students in $[\ldots AB]$ lose a candy piece. This may change the deficiency of each string and so contribute to a change to the index.

The following table outlines all possibilities and the change to the index the two strings $[\ldots A]$ and $[\ldots AB]$ together make. (Bettina possess at least two pieces of candy.)

$[\ldots A]$	$[\ldots AB]$	Possible?	Change to Index
Balanced	Balanced	NO	
Balanced	Deficient	NO	
Balanced	Abundant	YES	$0 + 0 = 0$
Deficient	Balanced	YES	$(-1) + 1 = 0$
Deficient	Deficient	YES	$(-1) + (-1) = -2$
Deficient	Abundant	YES	$(-1) + 0 = -1$
Abundant	Balanced	NO	
Abundant	Deficient	NO	
Abundant	Abundant	YES	$0 + 0 = 0$

In the first two strings we examined, Bettina alone and then all students but Bettina, together have a negative contribution to the index. We have just shown that all other strings, if they contribute to a change in the index make it smaller. Thus the index decreases each time a student performs a sharing action.

This establishes the claim, and thus proves our third observation.   □

This candy-sharing game is a special example of a parallel chip-firing game on a finite graph. (See [BITAR and GOLES], for example.)

After I sent out this candy-sharing research challenge, Scott Kominers of Harvard University showed me a paper he and his brother Paul Kominers had recently written (see [P. KOMINERS & S. KOMINERS]). In it they prove

Theorem. *If G is a finite, undirected, connected graph with v vertices and e edges, then any parallel chip-firing game on G with at least $4e - v$ chips eventually settles to a stable distribution of chips.*

A group of N students sitting in a circle can be viewed as a cyclic graph with N vertices and N edges. Thus, the Kominers brothers have proved

> *If N students are given $3N$ or more pieces of candy, the game they play will eventually settle into a fixed distribution of candy.*

Analysis of games with less than $3N$ pieces of candy is still open.

Challenge. Prove that, in the candy sharing game with N students, if the game does not terminate then each student fires infinitely often.

Challenge. Prove that in the candy sharing game with N students, if the game terminates, at least one student never fired at all.

We now have one final result:

Observation 4. *Suppose $k < N$ pieces of candy are distributed among N students, and suppose that the candy sharing game terminates when students share candy simultaneously at each blow of the whistle.*

Suppose the students decide to repeat the game (with the same initial distribution of candy) but each time sharing candy one student at a time in some non-specified order. Then:

i) *Each game again terminates ending with the same final distribution of candy (no matter the order in which students take turns sharing)*

ii) *The total number of times each student fires before the game terminates is constant.*

The proof of Observation 3 shows that with less candy than students, games are sure to terminate even if students take turns sharing one at a time.

Number the students 1 through N. We can record the play of a game as a sequence composed of these numbers. For example, $1 - 6 - 3 - 1 - 7 - \cdots$ reads "student 1 fired, then student 6 fired, then student 3 fired, then student 1 fired (again), then student 7 fired," and so on.

Suppose α and β are two sequences recording two different terminating games on the same initial distribution of candy. Let α_i be the number of times student i fires ($1 \leq i \leq N$) in the sequence α and define β_i similarly. We want to show that $\alpha_i = \beta_i$.

Suppose to the contrary that some value α_i is greater than its counterpart β_i (the case that β_i is greater than α_i is handled similarly) and imagine watching the plays of the games α and β simultaneously. There will be a first student who fires one more time in the game α than her counterpart ever fires in the entire game of β. Suppose this is student m. Then we can say that the sequence α has the form

$$\alpha = \alpha' - m - \alpha''$$

with initial part α' containing β_m appearances of m and no more than β_i appearances of i for all $i \neq m$.

That the sequence α exists means that if students play the partial sequence of moves given by α', then student m will be able to fire at the end of that play, that is, $\alpha' - m$ is legal. But $\beta - m$ would be legal too! In the play of β, m's neighbors may have fired more times than in α' and given m more candy. So if m is able to fire after the play of α', she is certainly able to fire after the play of β. This contradicts that β is a terminating game: no one should be able to make a play at the completion of β.

This establishes part ii) of Observation 4.

Now that we know that in each game played with the same initial distribution of candy the number of times each individual player fires is the same, it follows that the final distribution of candy is the same. We know how much candy player i begins with, we know how many pieces she receives from her neighbors (this is the sum of the number of times they each fire, fixed) and we know how many times player i will fire (also fixed). This uniquely specifies the count of candy she possesses at the completion of the game.

Challenge. Extend Observation 4 to the case of $k = N$ pieces of candy. Specifically, prove for an initial distribution of candy

i)   If one pattern for sharing the candy fails to terminate, then all sharing patterns will go on indefinitely.

ii) If one pattern for sharing the candy terminates, then all sharing patterns terminate (and the total number of times each student fires in a sharing pattern is fixed).

See [BJÖRNER, LOVÁSZ and SHOR] for a more general analysis on finite connected graphs.

References

[BITAR and GOLES] Bitar, J., and Goles, E., Parallel chip firing games on graphs, *Theoretical Computer Science,* **92** (1992), 291–300.

[BJÖRNER, LOVÁSZ and SHOR] Björner, A., Lovász, L., Shor, P.W., Chip-firing games on graphs, *European Journal of Combinatorics*, **12** (1991), 283–291.

[KOMINERS and KOMINERS] Kominers, P., and Kominers, S., A constant bound for the periods of parallel chip-firing games with many chips, To appear. (http://dx.doi.org/10.1007/s00013-010-0129-x) (Earlier version: "Candy-passing games on general graphs, II: http://arXiv.org/abs/0807.4655).

Appendix **VII**
Bending Buffon's Needle

In 1733 French naturalist and mathematician Georges-Louis Leclerc, Comte de Buffon (1707–1788) asked

If a needle one inch long is tossed onto a wooden floor with slats one inch wide, what is the probability that the needle will land crossing a line between two slats?

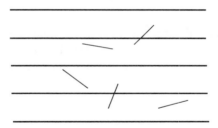

Forty-four years later, using integral calculus, Buffon published the answer: $\frac{2}{\pi}$, about 63.66%. (We'll confirm this later.)

Comment. We could, in principle, use this result to approximate the value of pi by dropping a needle onto a diagram of lines a large number of times and counting the crossings that occur. In practice, however, this turns out to be a very inefficient procedure: typically millions of tosses are needed to obtain a result accurate to just a few decimal places!

I recently asked followers of the St. Mark's Institute of Mathematics to ponder the following variation of the Buffon Needle Problem:

WOODEN FLOORS AND BENT NEEDLES

I have a wooden floor composed of uniform wooden slats each one inch wide making a pattern of parallel lines spaced one inch apart. I have a wire needle one-inch long and I am about to throw it on to the floor. But before I do so, I shall bend it.

a) **[WARM-UP]** *Suppose I bend the one-inch wire needle into a circle and throw it on to the floor. What is the probability that it lands touching or crossing a gap between two wooden slats?*

b) *Suppose instead I bend the needle into a square, each side one-quarter of an inch long, and throw it onto the floor. What is the probability that it lands touching or crossing a gap between two wooden slats?*

c) *Suppose instead I bend the needle into an equilateral triangle before throwing it. What is the probability that it lands touching or crossing a gap between two wooden slats?*

Comment. *Parts b) and c) are hard, unless one can come up with a philosophical reason as to why their answers must be. . .*

The first problem, part a), is straightforward: a ring of perimeter 1 has diameter $D = \frac{1}{\pi}$ and the probability of it crossing a line is $\frac{D}{1} = \frac{1}{\pi}$.

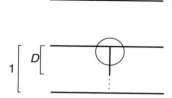

The remaining two questions turned out to be hard, though a number of Institute followers made valiant attempts at answering them. Several developed a sufficiently deep intuitive understanding of the problem to suspect that all three answers have to be the same (as hinted at in the final comment). This is indeed the case.

Bending the Needle

Let's rephrase Buffon's original result (a straight one-inch needle thrown against parallel one-inch slats) in terms of expected value:

When tossed on the set of parallel slats, the needle can cross either 0 or 1 lines. The expected number of crossings E for a random toss of the needle is:

$$E = 0 \cdot P \text{ (crosses zero times)} + 1 \cdot P \text{ (crosses once)}$$
$$= 0 + 1 \cdot \frac{2}{\pi}$$
$$= \frac{2}{\pi}.$$

(The probability that the needle will land exactly along a line is zero, as is the probability it will intersect two different lines. These situations are thus safely ignored.)

Now imagine two needles, needle A of length a (for $0 < a < 1$) and needle B of length $b = 1 - a$, tossed onto the same one-inch slats. Let $E(AB)$ denote the expected number of crossings from the pair of two needles together and let's write $A = i$ (or $(B = i)$ if needle A (or B) crosses a line i times (with $i = 0$ or 1). We have:

$$E(AB) = 0 \cdot P\,(A = 0 \text{ and } B = 0) + 1 \cdot P\,(A = 1 \text{ and } B = 0)$$
$$+ 1 \cdot P\,(A = 0 \text{ and } B = 1) + 2 \cdot P\,(A = 1 \text{ and } B = 1)$$
$$= P\,(A = 1 \text{ and } B = 0) + P\,(A = 1 \text{ and } B = 1)$$
$$+ P\,(A = 0 \text{ and } B = 1) + P\,(A = 1 \text{ and } B = 1)$$
$$= P\,(A = 1) + P\,(B = 1)$$
$$= E(A) + E(B),$$

where $E(A)$ is the expected number of crossings of needle A and $E(B)$ the expected number of crossings of needle B.

Nothing in the calculation relied on knowing how needles A and B interact with one another. If the two needles are independent of each other, then $E(AB) = E(A) + E(B)$. If needle B is glued to the end of needle A at a fixed angle, then $E(AB) = E(A) + E(B)$. If the two needles are attached to make an X-shape, then $E(AB) = E(A) + E(B)$ still holds.

The argument applies to any (finite) number of needles attached to each other, or independent from one another. (Let "A" represent one needle with crossing values 0 or 1, and let "B" represent all remaining needles with crossing values $0, 1, 2, 3, \ldots$. Follow the same calculation.) Thus all needles, say one inch long, composed of a large number of short straight line segments produce the same expected number of crossings.

As any (sufficiently nice) curve can be approximated by straight line segments, taking this argument to the limit yields the same invariance of the expected number of crossings. We have

All needles of length 1 inch, no matter how they are bent, have the same expected number of crossings.

As a circular needle of perimeter 1 has probability $\frac{1}{\pi}$ of crossing the lines twice, we have

This expected number of crossings of all 1-inch needles is $\frac{2}{\pi}$.

Any convex shape of perimeter 1, such as a square or an equilateral triangle, crosses a line either 0 or 2 times. (The exceptional cases occur with probability zero.) Thus

$$\frac{2}{\pi} = 0 + 2 \cdot P\,(\text{crosses})$$

and so the probability of crossing is $\frac{1}{\pi}$ in both cases.

For a straight needle of length 1, we have

$$\frac{2}{\pi} = 0 + 1 \cdot P\,(\text{crosses})$$

and so the solution to Buffon's original problem is $P\,(\text{crosses}) = \frac{2}{\pi}$.

Comment. When Buffon's needle is bent into a curve, the Buffon needle problem is called "Buffon's noodle problem." (See [RAMALEY].)

Challenge. A floor is tiled with squares each one inch wide in the standard regular pattern. A needle of length one inch is tossed onto the floor. What is the expected number of crossings across the gaps between tiles?

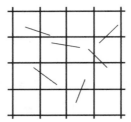

The Classic Approach to Buffon's Needle Problem

Suppose, once more, that a straight one-inch is tossed onto a set of parallel lines, one inch apart.

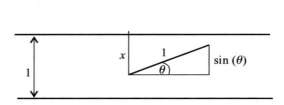

Let x be the distance of the lowest point of the needle from the parallel line above it, $0 \le x \le 1$, and θ the angle at which the needle lies as shown in the diagram, $0 \le \theta \le \pi$.

The needle crosses the line if $0 \le x \le \sin(\theta)$. So we ask: Of all the points (x, θ) with $0 \le x \le 1$ and $0 \le \theta \le \pi$, what are the chances that we select a point with $0 \le x \le \sin(\theta)$? This is a geometric problem.

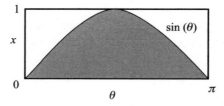

The chance of selecting a point in the shaded region shown is

$$P = \frac{\int_0^\pi \sin\theta \; d\theta}{\pi} = \frac{2}{\pi}.$$

Comment. We already knew the answer to the problem: $P = \frac{2}{\pi}$! This means we could have deduced without calculus that the area under the sine curve is two!

Challenge. What other areas under curves can we compute using Buffon's needles and noodles instead of calculus?

We end with a word of warning! We need to be careful when analyzing problems in probability theory. Here's an extraordinary example:

Two points are selected at random from the interval [0 , 1]. *What is the probability that they are no more than half a unit apart?*

Answer 1. Choosing two values x and y from [0, 1] is equivalent to choosing a point (x, y) from the unit square [0, 1] × [0, 1]. The two values will be less than a half unit apart, $|y - x| \leq \frac{1}{2}$, if the point (x, y) lands in the shaded region shown.

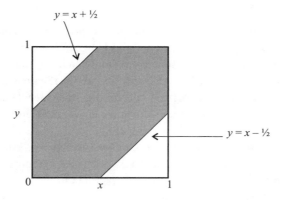

The probability we seek is thus $P = \frac{3}{4}$.

Answer 2. The point x will either lie in the left half [0, 0.5] or the right half (0.5, 1] of the unit interval. Let's flip a coin to decide which. (Heads for left, tails for right, say). The point y will either lie to the left of x or to the right of x. Let's decide this by a second flip of the coin.

There are four possibilities to consider, each with $1/4$ chance of occurring.

Case 1. $0 \leq x \leq 0.5$ and y is to the left of x.

Then x and y are certain to be no more than a half a unit apart.

Case 2. $0 \leq x \leq 0.5$ and y is to the right of x.

Then the probability that x and y are no more than half a unit apart is $\frac{2}{3}$.

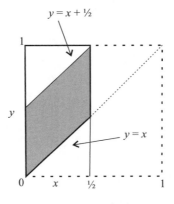

Case 3. $0.5 < x \le 1$ and y is to the left of x.

As in case 2, the probability that x and y are no more than half a unit apart is $\frac{2}{3}$.

Case 4. $0.5 < x \le 1$ and y is to the right of x.

Then x and y are certain to be no more than a half a unit apart.

The desired probability we seek is thus

$$P = \frac{1}{4} \times 1 + \frac{1}{4} \times \frac{2}{3} + \frac{1}{4} \times \frac{2}{3} + \frac{1}{4} \times 1 = \frac{5}{6}.$$

Both arguments are correct! So what is going on?

The issue is the use of the words "at random" in the problem statement. There are many different procedures for choosing points at random and they can lead to different results. In any problem in probability theory the problem statement should indicate the procedure in choosing random quantities. This is a subtle point that can lead to paradoxes if care is not taken.

So . . . Is the word "random" properly defined in Buffon's classic needle problem? Is it possible to analyze the problem in a different manner and obtain an answer different from $\frac{2}{\pi}$?

Challenge. Define an alternative procedure for choosing two points x and y at random from the interval $[0, 1]$ so that the probability that those two points are no more than half a unit apart is different from $\frac{3}{4}$ and from $\frac{5}{6}$. In fact, show that it is possible to define random procedures that produce *any* answer between 0 and 1 for this problem!

Reference

[RAMALEY] Ramaley, J. F., Buffon's noodle problem, *American Mathematical Monthly*, **76** (1969), 916–918.

Appendix **VIII**
On Separating Dots

The Intermediate Value Theorem shows that for any planar region (with a well-defined meaning of "area") there is a straight line that separates it into two parts with equal area. The Two-Pancake Theorem states that two planar regions can be simultaneously so divided with a single straight line cut. (See [TANTON] for details.) As we saw in Appendix III, for two continuous functions on the surface of a sphere there exist two antipodal points at which their values match.

In the Spring of 2010 students of the St. Mark's Institute research class explored discrete versions of these three classic results. Here is what they discovered.

RESULT 1: One Pancake Made Discrete

Suppose an even number of dots are scattered about a page. Then there is a straight line that

 i) *passes through no dot,*

 ii) *separates the dots into two groups of equal number on either side of the line.*

Proof. Let the number of dots be $2N$.

Sweep a horizontal line down from the top of the page across the dots towards the bottom of the page. Label the dots 1, 2, 3, ..., $2N$ according to the order in which the line encounters them. If the line encounters several dots simultaneously, label the numbers from left to right.

If the dots labeled N and $N + 1$ have different heights, then any horizontal line between them does the trick and separates the $2N$ dots.

If dots labeled N and $N + 1$ have the same height, then draw the horizontal line that passes through them and rotate it about the midpoint of the line segment connecting them. For a sufficiently small angle of rotation, the turned line will not pass through any of the remaining dots and so separates the set of $2N$ dots appropriately. $\qquad\square$

RESULT 2: Two Pancakes Made Discrete

Suppose $2N$ red dots and $2N$ blue dots are scattered about a page. Provided no three dots among them are collinear, there is a line that

i)   passes through no dot,

ii)   has N red dots and N blue dots on one side, and N red and N blue dots on the other.

Proof. Draw a circle that encloses the $4N$ dots. For each angle θ from the horizontal there is a directed line tangent to the circle as shown in Figure 1.

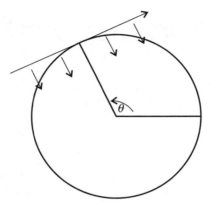

Figure 1

If we sweep the line across the $4N$ dots towards the center of the circle in a perpendicular direction, we can label the red dots $1, 2, 3, \ldots, 2N$ according to the scheme on the previous page and label the blue dots $1, 2, 3, \ldots, 2N$ similarly. Thus each angle θ produces a labeling scheme for each set of colored dots.

For each angle θ, we have a "band" of red line segments parallel to the directed tangent at that angle that lies between the red dots labeled N and $N + 1$, and a band of blue line segments that lie between the blue dots labeled N and $N + 1$. (A band could be one line in width if the dots labeled N and $N + 1$ are "simultaneous" for that labeling.)

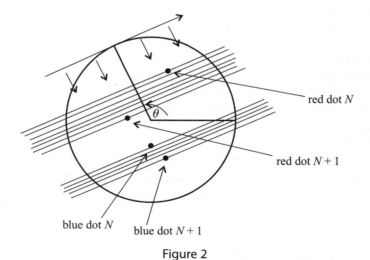

Figure 2

If the bands overlap, then there is a common line that passes between or on red dots N and $N + 1$ and is between or on blue dots and blue dots N and $N + 1$.

If the bands do not overlap, we change θ and argue that the width of the bands varies continuously as we move through different angles. (Experiment with a diagram containing just two and then just four red dots to see how the central band mutates as one changes the angle θ.) Actually, if for each red band we draw its central-most red line and for each blue band its central line, we can argue that the distance between the central lines varies continuously as we change θ.

In Figure 2 with angle θ, we have the band of red lines lying closer to the tangent line than the band of blue lines. For the angle $\theta + 180°$, we will have the same bands of lines, but the band of blue lines will be closer to the tangent line than the band of red lines. The central lines to the bands have swapped roles.

So as we vary the angle, there must be an angle between θ and $\theta + 180°$ where the central bands overlap and we have thus found a line that passes between or on red dots N and $N + 1$ and passes between or on blue dots N and $N + 1$.

Now that we have found a common line, we have three cases to consider.

1. No dots lie on the common line.

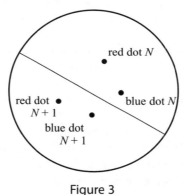

Figure 3

We're done!

2. One dot lies on the common line.

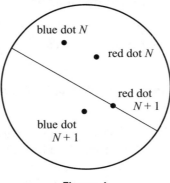

Figure 4

A slight shift of that line does the trick in all cases.

3.   Two dots lie on the common line.

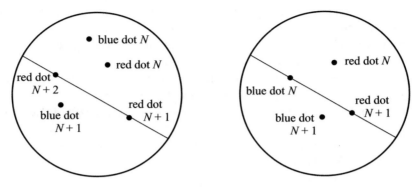

Figure 5

A small shift or rotation about the midpoint of the two dots on this line does the trick in all cases.
    More than two dots cannot be on the line because no three points are collinear.   □

Comment. If all $4N$ dots are collinear, then there certainly need not be a single separating line.
But is our condition "no three dots are collinear" excessively strong?

Challenge. Does result 2 extend three colors with $6N$ dots, $2N$ of each of three colors?

Challenge. Conduct an alternative approach to proving the theorem. Given $n \in \mathbb{N}$, expand each
dot to a colored disc of radius $\frac{1}{n}$. By the Two-Pancake Theorem there is a line that simultaneously
divides the total area of red into two and the total area of blue into two. Argue that in the limit as
$n \to \infty$ there is a limiting line that solves the problem.

Comment. It is perturbing that proving a discrete result requires properties of continuity. Surely
there is a more direct approach that doesn't require this big gun.

RESULT 3: Borsuk-Ulam Theorem Made Discrete

*Suppose $2N$ distinct red dots and $2N$ distinct blue dots are scattered about the surface of a sphere.
If no three dots among the $4N$ dots lie on the same great circle, there is at least one great circle that*

i)   *passes through no dot,*

ii) *has N red dots and N blue dots in one hemisphere defined by the circle, and N red and N blue
     dots in the other.*

This is a Borsuk-Ulam type result. For each point P on the surface of the sphere let $r(P)$ be the
number of red points that lie in the northern hemisphere defined by letting P be the north pole.
Define $b(P)$ to be the count of blue points in the northern hemisphere defined by P. Then result 3
claims that there exist two antipodal points P and P^* for which $r(P) = r(P^*)$ and $b(P) = b(P^*)$.

Proof. Define the values $r(P)$ and $b(P)$ as described above, but declare points that lie on the equator
of the northern hemisphere defined by P as each contributing a value one half to the count.

Let's focus on the red dots for the moment. If there were just one red dot on the sphere, the function r would have values as shown in Figure 6 for points on the sphere.

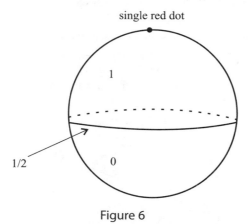

Figure 6

Let's call the equator of constant value half the "half-line" for the dot.

For a collection of $2N$ red dots on the sphere, the function r is a sum of $2N$ of these basic functions. (Figure 7 shows the values of r for points on the sphere for two red dots.) If P and P^* are antipodal points, then $r(P) + r(P^*) = 2N$.

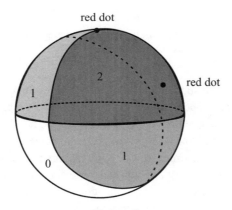

Figure 7

The half-lines subdivide the sphere into polygonal regions giving a tiling of the sphere. (Figure 7 is a tiling with four regions.) The function r has constant (integer) value on the interior of a region and its value differs by ± 1 on neighboring regions.

Because no three red dots are on the same great circle, no two red dots are antipodal and no three half-lines ever pass through the same point. Thus the vertex of each polygonal region of the tiling lies on two half-lines. The function r has integer value at each vertex and half-integer values at other boundary points.

On two neighboring regions that share an arc of common boundary, r has integer values that differ by one. Moreover, if r has integer value a at some interior point P of a polygon of the tiling, it has value $2N - a$ at its antipode P^*. Thus if we move along a path from P to P^* avoiding the vertices in the tiling, we will pass through a region of the tiling on which r has value N. (Its matching antipodal region has value N as well.)

Actually, *every* path from P to P^* either passes through an interior region of constant value N or through a vertex of value N.

Look at all the vertices and regions on which r has value N. There are only finitely many such regions and vertices (with antipodal counterparts), and they must piece together so that every path from P and P^* passes through one of them. We conclude that there is a pair of antipodal points Q and Q^* and a path between them on which r has constant value N.

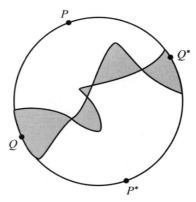

Thus we have established

> *There is a path from one point on the sphere to its antipode within one hemisphere (and back again within the opposite hemisphere) on which r has constant value N.*

The same is true for b.

It is not difficult to see that two such paths, in our case one for the red dots and one for the blue dots, must intersect at a pair of antipodal points. Let H be a point of intersection. The circle that is the equator of the northern hemisphere defined by H separates the red dots and it separates the blue dots. There still may be a problem: it itself might pass through some dots.

To finish the proof, we need to examine how to adjust things if this is indeed the case.

This circle won't pass through a single red dot (and some or no blue dots) as $r(H)$ has an integer value, namely N. Nor will it pass through a single blue dot for the same reason. As no three dots lie on the same great circle, the only possibility is that the circle passes through two dots of the same color. As there are only a finite number of dots on the sphere, a rotation of the great circle about the midpoint (and its antipode) of the arc that connects the two dots it passes through will fix the problem.

We are done!   □

Challenge. For each $n \in \mathbb{N}$ replace each dot on the sphere with a small disc of radius $\frac{1}{n}$. What does the (continuous) Borsuk-Ulam theorem say about finding a great circle that divides the total area of the red discs and the total area of the discs each in half? Can we take this result to the limit?

Challenge. Prove result 2 without using continuity!

Index of Topics

Classic Theorems Proved

About the Author

Born in Adelaide, Australia, James Tanton received his PhD from Princeton University in 1994 and worked at the college level a number of years before developing a deep interest in secondary mathematics education. He is committed to sharing the beauty of the subject to all, especially at the formative teenage years. James is the founding director of the St. Mark's Institute of Mathematics, an outreach program designed to promote joyful and effective mathematics education. He is a full-time high school teacher at St. Mark's School, Southborough, MA, and he conducts mathematics graduate courses for teachers through Northeastern University's School of Education.

James is the author of *SOLVE THIS: MATH ACTIVITIES FOR STUDENTS AND CLUBS* (MAA, 2001), *THE ENCYCLOPEDIA OF MATHEMATICS* (Facts on File, 2005), and ten self-published texts. He is the 2001 and the 2002 recipient of the Trevor Evans Award, the 2005 recipient of the Beckenbach Book Prize, the 2006 recipient of the Kidder Faculty Prize at St. Mark's School and a 2010 recipient of the Raytheon Math Hero Award for excellence in school teaching. He was also just nominated for a Presidential Teaching Award.

James continues to publish research and expository articles, and through his St. Mark's Institute of Mathematics research classes has helped high-school students pursue research projects and publish their results too.